Problem Solvers

Edited by L. Marder
Senior Lecturer in Mathematics, University of Southampton

No. 11

Statistics 1

Problem Solvers

1. ORDINARY DIFFERENTIAL EQUATIONS—J. Heading
2. CALCULUS OF SEVERAL VARIABLES—L. Marder
3. VECTOR ALGEBRA—L. Marder
4. ANALYTICAL MECHANICS—D. F. Lawden
5. CALCULUS OF ONE VARIABLE—K. Hirst
6. COMPLEX NUMBERS—J. Williams
7. VECTOR FIELDS—L. Marder
8. MATRICES AND VECTOR SPACES—F. Brickell
9. CALCULUS OF VARIATIONS—J. W. Craggs
10. LAPLACE TRANSFORMS—J. Williams
11. STATISTICS I—A. K. Shahani and P. K. Nandi

Statistics 1

Probability and Probability Distributions

A. K. SHAHANI and P. K. NANDI

University of Southampton and University of Bath

LONDON · GEORGE ALLEN & UNWIN LTD

RUSKIN HOUSE MUSEUM STREET

First published 1973

ISBN 0 04 519010 0 *paper*
0 04 519009 7 *hardback*

Set in 10 on 12 pt 'Monophoto' Times Mathematics Series 569
Printed in Great Britain by Page Bros (Norwich) Ltd., Norwich

Preface

Solving problems is of particular importance in a study of statistics. In this book we cover probability, random variable, and some particular distributions including binomial, Poisson and normal distributions. Many of the problems give a flavour of the practical applications of statistics. We have included a few standard results as problems, as well as some 'classical' problems. We have given a summary of theoretical ideas with definitions rather than proofs in mind. With judicious selection, this book should be of help to a variety of students.

We thank Professor M. H. Quenouille for reading a draft manuscript and suggesting many improvements. We thank Barry Blight for suggesting some problems, Les Marder for his helpful comments, and Dorothy Gauron and Eileen Richmond for looking after the typing.

A.K.S.
P.K.N.

Contents

Chapter 1

Probability

1.1 Introduction Words such as 'likely', 'may be', 'possible', and 'probable' are used quite frequently by many of us to express the idea that the truth of the matter at hand is in doubt. Mathematicians have formalized this idea into the concept of probability.

It is convenient to think about an experiment or a trial in which a defined event, A, may or may not occur. We take $P(A)$ to be the probability of A and note that

$$0 \leqslant P(A) \leqslant 1,$$

$$P(A) = 0 \quad \text{if} \quad A \text{ cannot occur,}$$

$$P(A) = 1 \quad \text{if} \quad A \text{ is certain to occur.}$$

The early development of the theory of probability was greatly influenced by the intensity with which European noblemen gambled with dice and cards. One gets a sense of this intensity from, for example, a problem given to Galileo (1564–1642). The problem was:

> 'When three dice are thrown, the numbers 9 and 10 can each be obtained in six different ways. Actual experience shows that 10 appears more often than 9. How can this discrepancy be accounted for?'

Galileo argued that the three dice are different and thus the total number of different cases for the three dice is $6 \times 6 \times 6$ or 216. The number 9 can in fact occur in 25 cases and there are 27 cases for the number 10. Thus the correct analysis bears out the practical experience, and we would say that the probability of 9 is 25/216 and that of 10 is 27/216. (What an intensity of gambling there must have been for the detection of this difference between an event with a probability 25/216 and an event with a probability 27/216!) The main line of this European development continued with the famous correspondence between Pascal (1623–1662) and Fermat (1601–1665), the work of Huygens (1629–1695), *Ars Conjectandi* by J. Bernoulli (1654–1705), *The Doctrine of Chances* by de Moivre (1667–1754) and the various important contributions by Laplace (1749–1827).

1.2 Some Basic Ideas

1.2.1 Equally likely outcomes In games with dice and cards, the notion of equally likely outcomes is a natural one. For example, in one throw of a regular die each of the six faces will be considered to be equally likely to come up. This notion of equally likely outcomes is used in the following rule for computing probabilities.

> If a trial can result in any one of n different, equally likely outcomes, and if exactly m of these outcomes mean the occurrence of an event A, then $P(A) = m/n$.

Note that as a definition of probability, the above rule is intellectually unsatisfactory, for it uses the concept of probability in the notion of equally likely events in order to define probability. Also, this rule covers the case of equally likely events only.

Problem 1.1 The ten digits 0 to 9 are stamped on ten discs, there being one digit on a disc, and the discs are thoroughly mixed in a box. Peter draws one disc at random; find the probability that the disc has an odd digit on it.

Solution. This trial can result in 10 different outcomes, and from the description of the trial it seems reasonable to suppose that these outcomes are equally likely. There are 5 ways in which an odd digit can occur. Thus the required probability is 5/10 or 0·5. □

Problem 1.2 A batch of components consists of d defective and $N-d$ good components. Find the probability that m randomly selected components will have exactly k defective components.

Solution. The number of combinations of n elements taken r at a time is

$$\binom{n}{r} = \frac{n!}{r!(n-r)!}.$$

The k defectives must come from the d defectives of the batch, and thus there are $\binom{d}{k}$ ways of getting the k defectives. There are $\binom{N-d}{m-k}$ getting $m-k$ good components. Thus there are $\binom{d}{k}\times\binom{N-d}{m-k}$ ways of getting exactly k defective components in the m selected components. The total number of ways of selecting the m components is $\binom{N}{m}$. The required probability is

$$\binom{d}{k}\binom{N-d}{m-k}\Big/\binom{N}{m}.$$ □

Problem 1.3 Peter and Paul each choose an integer from 1 to n (both inclusive) at random. For $n = 5$, find the probability that the total of the two integers chosen is: (a) even, (b) odd, and (c) greater than 4. For general n, find the probability that the two integers chosen differ by more than k, where $0 \leqslant k < n$.

Solution. For $n = 5$, Peter and Paul generate 25 possible outcomes. The number of outcomes for which the total is even, odd, and less than or equal to 4 is 13, 12, and 6 respectively. Thus the answers are: (a) $\frac{13}{25}$, (b) $\frac{12}{25}$, (c) $1-\frac{6}{25} = \frac{19}{25}$. Note that $\frac{13}{25}+\frac{12}{25} = 1$.

For general n, Peter and Paul generate n^2 possible outcomes and we are interested in the event $|x-y| > k$, where x, y are the chosen numbers. Consider the case of $(x-y) > k$. Now the smallest value that y can take is 1 so that the smallest allowable value for x is $k+2$. For a given x in $(k+2, n)$, y can take any one of the values $1, 2, \ldots, x-k-1$. Thus the total number of ways in which $(x-y) > k$ is

$$\sum_{x=k+2}^{n} (x-k-1) = \tfrac{1}{2}(n-k-1)(n-k).$$

Similarly there are $\frac{1}{2}(n-k-1)(n-k)$ ways in which $(x-y) < k$. Thus the required probability is $(n-k-1)(n-k)/n^2$. □

Problem 1.4 Find the probability of (a) at least one six in n throws of a die, (b) at least one double six in n throws of two dice. Hence show that there is an advantage in betting for one six in 4 throws of a single die and for double six in 25 throws of two dice.

Solution. There are 6^n possible outcomes for n throws of a single die. It is easier to count the number of ways in which six cannot occur, and from this deduce the number of ways in which six will occur at least once. For each of the n throws there are 5 ways in which six will not occur, so there are $6^n - 5^n$ ways in which six will occur at least once in n tosses. Thus the required probability for (a) is $(6^n - 5^n)/6^n = 1-(\frac{5}{6})^n$. Similarly the answer for (b) is $1-(\frac{35}{36})^n$.

Now $1-(\frac{5}{6})^4 = 0{\cdot}52$ and $1-(\frac{35}{36})^{25} = 0{\cdot}51$. Hence these bets would be advantageous, since $P > 0{\cdot}5$.

Comment. This is one of the problems solved by Pascal in response to a request from a French gambler Chevalier de Méré. □

1.2.2 Two other ideas Another rule for computing probabilities has evolved from practical experience with the relative frequency of the occurrence of an event in an extended series of trials. If $p(n)$ is the proportion of heads in n tosses of a coin, one feels that $p(n)$ will fluctuate in the neighbourhood of $\frac{1}{2}$ for large n. Practical experience supports this feeling. In general, if $p(n)$ is the proportion of times that an event A occurs in n repeated trials, the probability of A is taken to be

$$P(A) = \lim_{n \to \infty} p(n)$$

This meaning of probability as a long-term relative frequency increases the scope of applications beyond that of the 'equally likely' situations. The limit involved in this rule is not the usual mathematical limit, for the fluctuations in $p(n)$ cannot be guaranteed to be within some bounds for all values of n beyond a certain point. Note that this rule applies to trials that can be repeated and as such it does not cover events like 'there is a large quantity of oil under St. Paul's Cathedral' or 'Peter is guilty as charged'.

Another idea about probability is that it is a measure of a personal feeling about the truth of the matter at hand. The required probabilities are personally evolved through available knowledge and experience without necessarily using any further rules. A surgeon may, for example, sum up the prospects for a potential patient by telling him that the proposed operation has a better than 50–50 chance.

1.2.3 An axiomatic approach From a purely mathematical point of view a theory of probability, like any other mathematical theory, can be built up from some defined properties or axioms. In a purely mathematical theory it is not essential that probability should have any particular meaning. However, any set of axioms that is chosen for developing the theory should in some sense be acceptable. A satisfactory theory of probability has developed from about 1930 and the axioms used are motivated by the early work with equally likely cases and relative frequency. The standard notation of *sets* is useful. The set of all possible outcomes of an experiment, E, is called the *sample space*, and we shall denote this set by S. In carrying out E, we may be interested in events $A, B, C, \ldots$ which are subsets of S. Using the standard notation of sets, we note that:

1. $\bar{A}$, the *complement* of A, is the event that A does not happen.
2. $A \cup B$, the *union* of A and B, is the event that at least one of A or B happens.
3. $\bigcup_{i=1}^{n} A_i$, the union of sets $A_1, A_2, \ldots, A_n$, is the event that at least one of $A_1, A_2, \ldots, A_n$ happens.
4. $A \cap B$, the *intersection* of A and B, is the event that both A and B happen.

5. $\bigcap_{i=1}^{n} A_i$, the intersection of $A_1, A_2, \ldots, A_n$, is the event that all $A_1, A_2, \ldots, A_n$ happen.

6. The *empty set* is denoted by $\varnothing$, so that $A \cap B = \varnothing$ says that the sets A and B are disjoint. The events A and B are said to be *mutually exclusive.*

7. $\overline{(A \cup B)} = \bar{A} \cap \bar{B}$.

8. $\overline{(A \cap B)} = \bar{A} \cup \bar{B}$.

Given a sample space S, we associate a real number $P(A)$ with each event A and call $P(A)$ the probability of A. This probability satisfies the following properties.

1. $0 \leqslant P(A) \leqslant 1$.
2. $P(S) = 1$.
3. If A and B are mutually exclusive, $P(A \cup B) = P(A) + P(B)$.

3a. If A_i, $i = 1, 2, \ldots$, are pairwise mutually exclusive events then

$$P\left(\bigcup_{i=1}^{\infty} A_i\right) = P(A_1) + P(A_2) + \ldots + P(A_n) + \ldots .$$

1.3 Conditional Probability, Independence, and Two Useful Theorems Let S be the sample space corresponding to an experiment E and let A, B be two events with probabilities $P(A)$, $P(B)$ respectively. We denote the conditional probability of A given that B has happened by $P(A/B)$, where

$$P(A/B) = P(A \cap B)/P(B)$$

It follows from this definition that

$$P(A \cap B) = P(A)\,P(B/A) = P(B)\,P(A/B).$$

Note that if $P(A \cap B) = \varnothing$, that is if A and B are mutually exclusive, then $P(A/B) = 0$ and $P(B/A) = 0$.

We define A and B to be *independent* if and only if $P(A \cap B) = P(A)P(B)$. Note that for independent events A, B we have $P(A/B) = P(A)$ and $P(B/A) = P(B)$, which expresses the intuitive idea that independence of A and B means that the probability of any one of them is not influenced by the knowledge that the other has occurred. Given 3 events, A, B and C we could have pairwise independence, but not necessarily a general or mutual independence. We say that A, B, C are mutually independent if and only if all the following conditions hold.

$$P(A \cap B) = P(A)P(B) \qquad P(A \cap C) = P(A)P(C)$$
$$P(B \cap C) = P(B)P(C) \qquad P(A \cap B \cap C) = P(A)P(B)P(C).$$

This notion of mutual independence can now be generalized in an obvious manner to n events.

Theorem 1.1 If A and B are any two events, then

$$P(A \cup B) = P(A) + P(B) - P(A \cap B).$$

This result can be easily extended to the case of 3 events A, B and C. We have

$$P(A \cup B \cup C) = P(A) + P(B) + P(C) - P(A \cap B) - P(A \cap C) - P(B \cap C) + P(A \cap B \cap C).$$

For the case of n events A_i, we have

$$P\left(\bigcup_{i=1}^{n} A_i\right) = \sum_{i=1}^{n} P(A_i) - \sum_{i<j}\sum P(A_i \cap A_j) + \sum_{i<j<k}\sum\sum P(A_i \cap A_j \cap A_k) - \ldots + (-1)^{n-1} P\left(\bigcap_{i=1}^{n} A_i\right)$$

Theorem 1.2 If an experiment E can result in one and only one of the events B_i, $i = 1, 2, \ldots, n$, that is if B_i are such that. (a) $B_i \cap B_j = \varnothing$ for all $i \neq j$, (b) $\bigcup_{i=1}^{n} B_i = S$, and if A is a subset of S, then

$$P(B_i/A) = P(A/B_i)P(B_i) \Big/ \sum_{j=1}^{n} P(A/B_j)P(B_j), \qquad i = 1, 2, \ldots, n.$$

This important theorem is known as *Bayes' Theorem.*

Problem 1.5 A coin is tossed an even number of times. Find the probability of getting an equal number of heads and tails. Examine this probability as the number of tosses is increased.

Solution. It is reasonable to suppose that the probability of a head at any particular toss is $\frac{1}{2}$, independently of the outcome of the other tosses. Given this independence, we can say, for example, that the probability of two heads in two tosses is $\frac{1}{2} \times \frac{1}{2}$. In general, the probability of any particular sequence of heads and tails in $2n$ tosses is $(\frac{1}{2})^{2n}$, including the case of n heads and n tails. Now the number of different ways in which we can get n heads and n tails in $2n$ tosses is $\binom{2n}{n}$, and of course each of these different ways has a probability of $(\frac{1}{2})^{2n}$. Thus by property 3 given on page 5 the required probability is $P = \binom{2n}{n} (\frac{1}{2})^{2n}$. Now Stirling's approximation for $n!$ is $n! = \sqrt{(2\pi n)}\, n^n e^{-n}$. With this approximation, P reduces to $1/\sqrt{(\pi n)}$ which decreases with n. □

Problem 1.6 A system consists of three components A, B, C and the system will fail to operate if any one, or more, of the components fail. Each of the three components has, independently, a probability p of failing. (a) Find the probability that the system will fail. (b) Investigate the advantage of one extra B component which would mean that the system will fail due to the B component if and only if both the B components fail. The two B components would also fail independently each with probability p.(c) How many such B components would be needed in order to make the probability of the system failing at most $2p+p^2$?

Solution. (a)
$$\begin{aligned} P(\text{system fails}) &= 1-P(\text{system does not fail}) \\ &= 1-P(\text{all three components function}) \\ &= 1-(1-p)^3 = P_1. \end{aligned}$$

(b) With one extra B component, we have

$$P(\text{failure due to } B \text{ component}) = p^2.$$

Therefore

$$P(\text{system fails} = 1-(1-p)^2(1-p^2) = P_2.$$

Now $P_1-P_2 = p(1-p)^3 \geqslant 0$. Thus the extra B component reduces the probability that the system will fail.

(c) With n components of type B, we have:

$$P(\text{system fails}) = 1-(1-p)^2(1-p^n) = P_n.$$

With $P_n \leqslant 2p+p^2$,

$$n \geqslant 2+[\log 2-2\log(1-p)]/\log p$$

We would round up to get the smallest number of the necessary B components. □

Problem 1.7 Production of a component is scheduled on a daily basis and one of two independent machines M_1 and M_2 is chosen for the production. Probability that M_1 will be chosen is 0·6 and the probability that M_2 will be chosen is 0·4. Machines M_1, M_2 produce 2%, 5% defective components respectively. From one day's production a component is chosen at random and is found to be defective. Compute the probability that it was produced by M_1.

Solution. Let D denote the production of a defective component. Now

$$P(M_1 \cap D) = P(M_1)P(D/M_1) = 0{\cdot}6\times 0{\cdot}02 = 0{\cdot}012$$
$$P(M_2 \cap D) = 0{\cdot}4\times 0{\cdot}05 = 0{\cdot}020$$

Hence

$$P(D) = 0{\cdot}032 \quad \text{and} \quad P(M_1/D) = 0{\cdot}012/0{\cdot}032 = \tfrac{3}{8}.$$ □

Problem 1.8 Peter has three coins in his pocket. One of the coins is unbiased, that is $P(\text{heads}) = 0{\cdot}5$. The other two coins are biased with $P(\text{heads}) = 0{\cdot}4$ and $0{\cdot}3$. Peter selects one of the three coins at random, tosses it 3 times and observes 2 heads. Find the probability that the unbiased coin was selected.

Solution. Let B_1, B_2 and B_3 be the events that the selected coin has $P(\text{heads}) = 0{\cdot}3, 0{\cdot}4$ and $0{\cdot}5$ respectively. Let A be the event that 2 heads in 3 tosses are observed. Now

$$P(B_1) = P(B_2) = P(B_3) = \tfrac{1}{3}.$$

$$P(A/B_1) = \binom{3}{2}(0{\cdot}3)^2(0{\cdot}7), \quad P(A/B_2) = \binom{3}{2}(0{\cdot}4)^2(0{\cdot}6),$$

$$P(A/B_3) = \binom{3}{2}(0{\cdot}5)^3.$$

From Bayes' theorem, we have that

$$P(B_3/A) = P(A/B_3)P(B_3)\Big/\sum_{i=1}^{3} P(A/B_i)P(B_i)$$

$$= \binom{3}{2}(0{\cdot}5)^3\left(\frac{1}{3}\right)\Big/0{\cdot}284 = 125/284$$ □

Problem 1.9 Paul uses distilled water supplied in standard bottles by Peter. Peter's method of producing the distilled water is such that a bottle can be one of three types, A, B, C with probabilities 0·8, 0·15, 0·05 respectively. Type A is pure distilled water, type B contains a small percentage of impurity and type C contains a large percentage of impurity. Paul tests each bottle before using it and his testing procedure will pass the types A, B, C with probability 1·0, 0·5, 0·1 respectively. What percentage of the bottles sent by Peter will be passed by Paul in the long run?

Solution. Let D be the event that a bottle is passed by Paul. Now D can occur in any one of three mutually exclusive ways, namely, $D \cap A$, $D \cap B$, $D \cap C$. Therefore

$$\begin{aligned} P(D) &= P(D \cap A) + P(D \cap B) + P(D \cap C) \\ &= P(A)P(D/A) + P(B)P(D/B) + P(C)P(D/C) \\ &= 0{\cdot}8 \times 1{\cdot}0 + 0{\cdot}15 \times 0{\cdot}5 + 0{\cdot}05 \times 0{\cdot}1 = 0{\cdot}88. \end{aligned}$$

Regarding probability as a long term relative frequency, we see that Paul will pass 88% of the the bottles sent by Peter in the long run. □

Problem 1.10 A mixed doubles tennis tournament attracts n husbands and their wives and the pairings for the tournament are done at random. Find the probability that none of the husbands is paired with his wife.

Solution. Let E_i represent the event that the ith husband is 'correctly' paired. We want:

$$P(\bar{E}_1 \cap \bar{E}_2 \cap \ldots \cap \bar{E}_n) = P(\overline{E_1 \cup E_2 \cup E_3 \ldots \cup E_n})$$
$$= 1 - P(E_1 \cup E_2 \ldots \cup E_n)$$
$$= 1 - \sum P(E_i) + \sum P(E_i \cap E_j)$$
$$- \sum P(E_i \cap E_j \cap E_k) + \ldots$$
$$= 1 - n\,P(E_i) + \binom{n}{2} P(E_i \cap E_j) - \ldots$$

Now $P(E_i) = 1/n$, $P(E_i \cap E_j) = 1/n(n-1)$ etc. Thus the required probability is

$$1 - 1/1! + 1/2! - \ldots + (-1)^n/n!$$ □

Problem 1.11 Independent events P, Q and R have probabilities p, q and r, respectively, of occurring. A is the event 'P or Q but not both' and B is the event 'Q or R but not both'. Find $P(A/B)$.

Solution. $A = (P \cap \bar{Q}) \cup (Q \cap \bar{P})$, $B = (Q \cap \bar{R}) \cup (R \cap \bar{Q})$, so that

$$A \cap B = (P \cap R \cap \bar{Q}) \cup (Q \cap \bar{P} \cap \bar{R}).$$

Hence

$$P(A/B) = \frac{P(A \cap B)}{P(B)} = \frac{pr(1-q) + q(1-p)(1-r)}{q(1-r) + r(1-q)}$$
$$= \frac{q(1-r) + p(r-q)}{r + q - 2rq}.$$ □

Problem 1.12 A sequence of n telegraph signals consisting of dots and dashes is such that the probability of any particular signal being a dot is $\frac{3}{7}$ and the probability of a dash is $\frac{4}{7}$. The signals are independent. Find the probability that the sequence will contain no consecutive dots or dashes.

Solution. We must have alternate dots and dashes. The nature of n, even or odd, has to be taken into account.

Even n: There should be $n/2$ dots and $n/2$ dashes and the sequence could start either with a dot or with a dash. Thus the required probability P is $P = 2(\frac{3}{7})^{n/2}(\frac{4}{7})^{n/2}$.

Odd n: Here $(n-1)$ is even and apart from the first signal which could be a dot or a dash, we should have $(n-1)/2$ dots and $(n-1)/2$ dashes. Thus

$$P = \tfrac{3}{7}(\tfrac{4}{7})^{(n-1)/2}(\tfrac{3}{7})^{(n-1)/2} + \tfrac{4}{7}(\tfrac{4}{7})^{(n-1)/2}(\tfrac{3}{7})^{(n-1)/2} = (\tfrac{4}{7})^{(n-1)/2}(\tfrac{3}{7})^{(n-1)/2}.$$ □

1.4 Recurrence Relations There are many problems which are not easy to solve by the methods of the previous sections. In some problems the event of interest can occur at different stages during a process, for example the experiment E in which a series of independent trials is continued until a defined event occurs. In this E, the event of interest can occur at trial number i, $i = 1, 2, 3, \ldots$. In many such problems, we can set up relationships between the probabilities of the event at the various stages and solve the resulting equations.

Problem 1.13 For the sequence of n telegraph signals of Problem 1.12, find the probability that the sequence will contain no consecutive dots.

Solution. Let P_n = probability that there are no consecutive dots in n signals. Suppose there are no consecutive dots in $(n-1)$ signals. Now the nth signal can be a dash in which case it does not matter whether the signal number $n-1$ was a dot or a dash. But if the nth signal is a dot we must have signal number $(n-1)$ equal to a dash, and of course there will have been no consecutive dots in the remaining $(n-2)$ signals. Thus we get

$$P_n = \tfrac{4}{7}P_{n-1} + \tfrac{3}{7}\times\tfrac{4}{7}P_{n-2},\ n \geqslant 2;\quad P_0 = 1;\quad P_1 = 1.$$

This is a difference equation of order 2 and the solution is $P_n = A\lambda_1^n + B\lambda_2^n$, where λ_1, λ_2 are the roots of $x^2 = \frac{4}{7}x + \frac{12}{49}$ and A, B have to be evaluated from the initial conditions. Thus

$$P_n = A(\tfrac{6}{7})^n + B(-\tfrac{2}{7})^n.$$

$P_0 = 1$ and $P_1 = 1$ lead to the evaluation of A, B and we get

$$P_n = (\tfrac{1}{7})^n \times \tfrac{1}{8}\{9\times 6^n - (-2)^n\}. \qquad \square$$

Problem 1.14 In a series of trials, each trial can result in 0 or 1, and the probability of 1 in a trial is p_0 if the previous trial had resulted in 0, and it is p_1 if the previous trial had resulted in 1. For the first trial, the probability of 1 is p. Find the probability that the nth trial will result in 1.

Solution. Let f_n be the required probability. Note that trial number $n-1$ results in either 0 or 1. Now

$$f_n = p_1 f_{n-1} + p_0(1-f_{n-1}),\ n > 1, \qquad f_1 = p.$$

This is a simple difference equation of order 1 and can be readily solved. We have

$$f_n = p_0 + (p_1-p_0)f_{n-1} = p_0 + p_0(p_1-p_0) + (p_1-p_0)^2 f_{n-2}.$$

Repeated substitution gives

$$\begin{aligned} f_n &= p_0[1+(p_1-p_0)+\ldots+(p_1-p_0)^{n-2}] + p(p_1-p_0)^{n-1} \\ &= p_0[1-(p_1-p_0)^{n-1}]/(1-p_1+p_0) + p(p_1-p_0)^{n-1} \end{aligned} \qquad \square$$

Problem 1.15 A rook is placed on a specified square on a chess board.

At each move it has an equal probability of moving to any one of the squares, other than the one currently occupied, in the same rank or file. Find the probability that it will return to, (a) the original rank, and (b) the original square at move number n.

Solution. (a) Let $P_n(i)$ be the probability that the rook is on rank i at the nth move. Now the rook can move to rank i from rank $j \neq i$ in one move with probability $\frac{1}{14}$ and it can stay in rank i in one move with probability $\frac{1}{2}$. Thus we get the recurrence relation:

$$\begin{aligned} P_n(i) &= \tfrac{1}{14}\{1-P_{n-1}(i)\}+\tfrac{1}{2}P_{n-1}(i) \\ &= \tfrac{1}{14}+\tfrac{3}{7}P_{n-1}(i) = \tfrac{1}{8}+(\tfrac{3}{7})^n\{P_0(i)-\tfrac{1}{8}\}. \end{aligned}$$

If the rook is placed on ith rank initially, we have $P_0(i) = 1$. Thus the required probability, $P_n = \frac{1}{8}+\frac{7}{8}(\frac{3}{7})^n$. Note that the probability that the rook returns to the original file at move n is also $\frac{1}{8}+\frac{7}{8}(\frac{3}{7})^n$.

(b) Let $Q_n(i,j)$ be the probability that the rook will be on square (i,j) at nth move. Clearly at move $n-1$, the rook must either be in rank i or file j but it must not be on square (i,j). From its position at move $n-1$, the rook has to move to square (i,j) in one move. Thus we have the recurrence relation:

$$Q_n(i,j) = \tfrac{1}{14}[P_{n-1}(i)-Q_{n-1}(i,j)]+\tfrac{1}{14}[P_{n-1}(j)-Q_{n-1}(i,j)]$$

With $Q_0(i,j) = 1$, we have the recurrence relation

$$Q_n = \tfrac{1}{7}[\tfrac{1}{8}+\tfrac{7}{8}(\tfrac{3}{7})^{n-1}]-\tfrac{1}{7}Q_{n-1}$$

This difference equation has the solution $Q_n = \frac{1}{64}+\frac{14}{64}(\frac{3}{7})^n+\alpha(-\frac{1}{7})^n$, where α depends on the initial conditions. Now $Q_0 = 1$ gives $\alpha = \frac{49}{64}$. Thus the solution is

$$Q_n = \tfrac{1}{64}[1+14(\tfrac{3}{7})^n+49(-\tfrac{1}{7})^n]. \qquad \square$$

1.5 Geometrical Probability The probability that a randomly selected point will be contained in a domain is proportional to the size of the domain. Thus if the total size of a domain is T, the probability that a randomly chosen point in this domain will be contained in a given sub-domain of size D is D/T. The selection process has to be carefully specified.

Problem 1.16 A point is selected at random on a straight line AE, of lenth $4l$. The straight line is divided into four segments, AB, BC, CD and DE, each of length l. Find the probability that the point will be in AB or CD.

Solution. We suppose that the selection process is such that the probability of a point being in a section of the line is proportional to length of the section. The size of the subdomain of interest is $2l$, thus the required probability is $2l/4l$ or $\frac{1}{2}$. □

Problem 1.17 In a test of the strength of a certain steel, a rigid bar of length L is subjected to a given strain. Experience has shown that 90% of the bars survive the test and 10% of the bars break at a random point. Find the probability that a bar will break at a point within $\frac{1}{4}L$ from any one of the two ends.

Solution. Let A be the event that a bar breaks, and let B be the event that a break occurs within $\frac{1}{4}L$ from anyone of the two ends. Now $P(A) = 0{\cdot}10$, $P(B/A) = \frac{2}{4}L/L = 0{\cdot}5$. The required probability is $P(A \cap B) = 0{\cdot}05$. □

Problem 1.18 A fixed diameter of a circle of radius r is AB. A point is chosen at random within the circle. Find the probability that the point is within an inscribed isosceles triangle with AB as one of the three sides.

Solution. The area of an inscribed isosceles triangle ABC, Fig. 1.1 is r^2; there are two such triangles. Thus the required probability P is $P = 2r^2/\pi r^2 = 2/\pi$. □

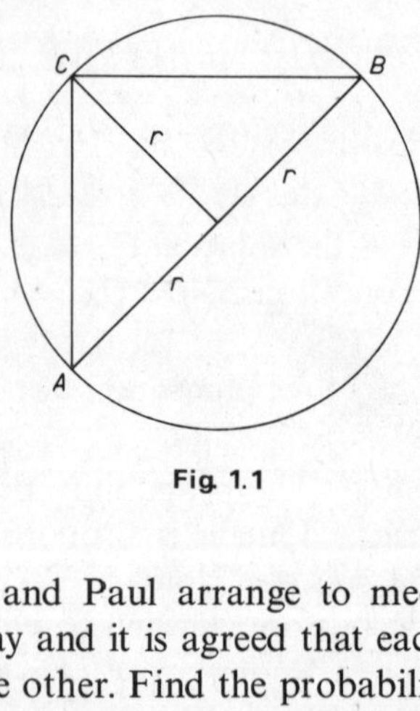

Fig. 1.1

Problem 1.19 Peter and Paul arrange to meet between 10 a.m. and 11 a.m. on a certain day and it is agreed that each should wait no longer than 10 minutes for the other. Find the probability that they will meet.

Solution. Let x, y be the times of arrival of Peter, Paul respectively. It is convenient to take 10 a.m. as the origin, and to measure time in minutes. Now Peter and Paul will meet if $x-10 \leqslant y \leqslant x+10$. Thus the required probability P is obtained by computing the ratio of the shaded to the total area from Fig. 1.2. We get

$$P = (3600 - 2 \times 1250)/3600 = \tfrac{11}{36}.$$ □

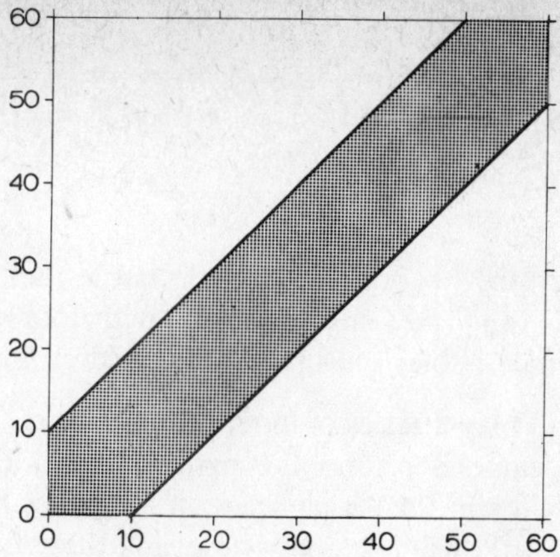

Fig. 1.2

Problem 1.20 Two points are selected at random on a straight line of length l. Find the probability that a triangle can be constructed from the resulting three sections.

Solution. Let x and y denote the position of the two points from one end of the line and suppose that $y \geqslant x$. The three sections thus have lengths x, $y-x$, $l-y$. For a triangle we must have

$$x+(y-x) > l-y \text{ or } y > l/2, \quad x+(l-y) > y-x \text{ or } (y-x) < l/2,$$
$$(l-y)+(y-x) > x \text{ or } x < l/2.$$

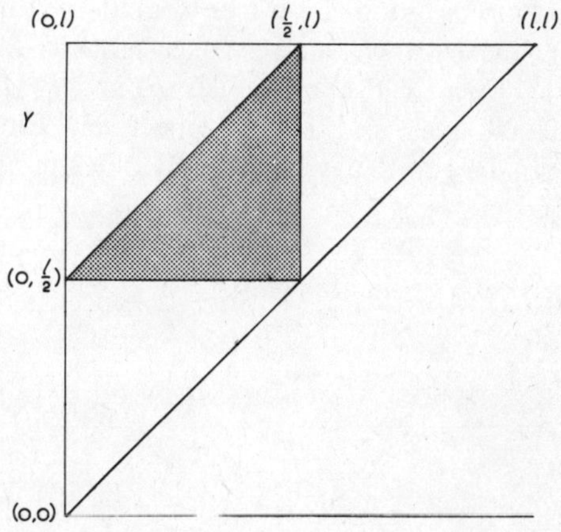

Fig. 1.3

The possible ranges for x, y are given by $0 \leqslant x \leqslant y$ and $0 \leqslant y \leqslant l$. Thus the required probability P is the ratio of the area of the shaded triangle to the larger triangle as shown in Fig. 1.3. We get $P = \frac{1}{8}l^2/\frac{1}{2}l^2 = \frac{1}{4}$. □

EXERCISES

1. In a large city the proportion of male babies is 0·5. Show that the probability that a randomly chosen family with four children will have had exactly two male babies out of the four children is not 0·5.

2. Peter and Paul play a series of independent games. The probability of Peter winning a game is p, and the probability of Paul winning it is $q = 1-p$. Peter will win the series if he wins r games before Paul wins s games, otherwise Peter will lose the series. Find the probability that Peter wins the series.

3. From a bag containing nine discs, numbered 1 to 9, three discs are drawn at random without replacement. Find the probability that the largest number obtained will exceed the smallest by at least 3.

4. A particle moves between k positions $A_j, j = 1, 2, \ldots, k$. At the starting time 0, the particle is in one of the A_j positions and at times $1, 2, 3 \ldots$ the particle leaves the position currently occupied and moves to one of the remaining $(k-1)$ positions, each of these $(k-1)$ positions being equally likely to be occupied. Find the probability that at time n the starting position will be occupied.

5. A device is said to be 'good' if it is free from all the three types of defect 'mechanical (M)', 'electrical (E)' and 'surface finish (F)'. It is known that $P(M)$, $P(E)$, $P(F)$, $P(M \cap E)$, $P(M \cap F)$, $P(E \cap F)$ and $P(M \cap E \cap F)$ are 0·05, 0·02, 0·10, 0·01, 0·008, 0·002 and 0·0005 respectively. Find the probability that a device (a) will be good, (b) will have any one, but not more than one, type of defect.

Chapter 2

Random Variable: Univariate Case

2.1 Definition We have considered the sample space S of an experiment E, where the elements s of S are the possible outcomes of E. These elements may or may not be numerical quantities. Often the outcome is a real number, or the outcome is converted into a real number. For example, the sample space associated with the tossing of a coin twice may be represented by:

$$S = \{HH, TH, HT, TT\}.$$

If now we are interested in the number of heads, we would assign the numbers 2, 1, 1, 0 to the four elements of S. In general we are often interested in assigning a real number to every outcome s; thus we are interested in a function X such that $x = X(s)$. This function X is called a random variable or a variate.

2.2 Discrete Random Variable If the number of possible values of X is finite or countably infinite, we say that X is a discrete random variable. For a discrete random variable we can write the possible values of X as $x_1, x_2, x_3, \ldots, x_i, \ldots, x_n, \ldots$ and this list would terminate in the finite case. We associate probabilities with x_i, and we use the notation

$$P(x_i) = P(X = x_i).$$

Note that (a) $0 \leqslant P(x_i) \leqslant 1, i = 1, 2, 3 \ldots$. (b) $\sum_{i=1}^{\infty} P(x_i) = 1$.

The function $P(x)$ can be thought of as showing how the total probability of 1 is distributed over the possible values of X. The collection of the values $\{x_i, P(x_i)\}$, $i = 1, 2, 3, \ldots$ is called the probability distribution of X.

Integral valued random variables are especially important and we shall often denote an integral-valued random variable by R. Typically R would take the value r, where $r = 0, 1, 2, 3, 4, \ldots$.

An important function is the cumulative distribution function $F(x)$ (c.d.f), sometimes called the distribution function, of a random variable X. The c.d.f. is defined to be

$$F(x) = P(X \leqslant x).$$

Note that $F(x) = \sum_i P(x_i)$, where the sum is taken for all i such that $x_i \leqslant x$. If we have an integral-valued random variable R which takes the values 0, 1, 2, 3,..., the c.d.f. can be written as $F(r) = \sum_{j=0}^{r} P(j)$.

Problem 2.1 A random variable R takes the value r with probability $k\alpha^r$, $r = 0, 1, 2, \ldots, l$. Find k and $P(R \leqslant l-1)$.

Solution. Let $P(r) = k\alpha^r$. Now $\sum_0^l P(r) = k(1-\alpha^{l+1})/(1-\alpha) = 1$. Thus $k = (1-\alpha)/(1-\alpha^{l+1})$ and

$$P(R \leqslant l-1) = \sum_0^{l-1} P(r) = \frac{1-\alpha^l}{1-\alpha^{l+1}} = 1-P(R = l). \qquad \square$$

2.3 Continuous Random Variable If X can take any value in an interval, say (a, b), the possible values of X are not countable and $P(x_i)$ ceases to be a useful function. In this case we say that X is a continuous random variable or variate. The c.d.f., $F(x) = P(X \leqslant x)$, is important and if it is differentiable at x, we define

$$f(x) = dF(x)/dx.$$

This function $f(x)$ is called the probability density function (p.d.f.) of X. If $f(x)$ exists, we take $f(x)dx$ to be the probability that X will take a value in a small interval which includes x. We may express this interpretation of $f(x)$ by writing

$$f(x)dx = P(x-dx/2 < x < x+dx/2).$$

In general we may take the possible values of X to be $-\infty < x < \infty$.

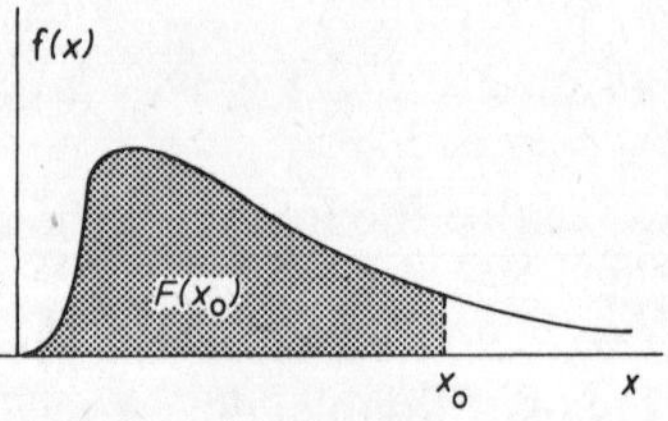

Fig. 2.1

Given $f(x)$ we can think of $F(x)$ as being the area under $f(x)$ as shown in Fig. 2.1. Note that:

(a) $f(x) \geqslant 0$.

(b) $F(x) = \int_{-\infty}^{x} f(t)\, dt$.

(c) $F(x_1) > F(x_2)$ if $x_1 > x_2$.

(d) $F(-\infty) = 0, F(\infty) = 1$.

(e) $P(c < x < d) = F(d)-F(c)$, $d > c$.

(f) $P(x) = 0$.

We say that x_p is the $(100p)$th percentile of X when $P(X \leqslant x_p) = p$. In particular $x_{0\cdot5}$ is called the median of X.

Another function, especially useful when the random variable T is the 'life time' ($\geqslant 0$) of an item, is $r(t)$ where $r(t) = f(t)/(1-F(t))$. Note that $r(t)dt$ would mean probability $(t < T \leqslant t + dt \mid T > t)$. This function $r(t)$ has been given several names, for example, force of mortality (actuarial usage), age specific failure rate, failure rate, hazard rate, etc. Note that:

$$r(t) = -d\{\log(1-F(t))\}/dt$$

$$1-F(t) = \exp[-\int_0^t r(u)\,du]$$

$$f(t) = r(t)\exp[-\int_0^t r(u)\,du].$$

This conditional function may also be defined for the discrete case and we could take $q(r) = P(r)/\sum_r^\infty P(j)$.

Problem 2.2 A random variable X, which takes values in $(0, a)$, has the distribution function kx^2, $0 < x < a$. Find the p.d.f. and $P(X > \frac{1}{2}a)$.

Solution. Let $F(x) = kx^2$, $F(a) = ka^2 = 1$ and therefore $k = 1/a^2$.

$$F'(x) = 2x/a^2 = f(x)$$

$$P(X > \tfrac{1}{2}a) = 1-F(\tfrac{1}{2}a) = \tfrac{3}{4}. \qquad \square$$

Problem 2.3 A random variable X has probability density function $f(x) = 2x$, $0 < x < 1$. Find the 25th percentile and the median of X.

Solution. Now $F(x) = \int_0^x 2t\,dt = x^2$ and $F(x) = 0\cdot25$ gives $x = 0\cdot5$. Thus $x_{0\cdot25} = 0\cdot5$. $F(x) = 0\cdot5$ gives $x_{0\cdot5} = \sqrt{0\cdot5}$.

Problem 2.4 Find the age specific failure rate when $f(t) = \lambda e^{-\lambda t}$, $t > 0$.

Solution. $1-F(t) = \int_t^\infty e^{-\lambda t}\,dt = e^{-\lambda t}$. Thus $r(t) = \lambda$.

Note that $r(t)$ is independent of t when $f(t) = \lambda e^{-\lambda t}$. A physical interpretation is that the random variable T corresponds to items which do not 'age'. $\square$

Problem 2.5 A random variable T has p.d.f. $f(t)$, $t \geqslant 0$. Its age specific failure rate $r(t) = t$. Find $f(t)$ and $P(T \leqslant t_0)$.

Solution. Now $\int_0^t r(u)\,du = \frac{1}{2}t^2$. Thus $F(t) = 1-e^{-t^2/2}$ so that $f(t) = F'(t) = te^{-t^2/2}$.

Also $P(T \leqslant t_0) = F(t_0) = 1-e^{-t_0^2/2}$. $\square$

Problem 2.6 A test of the strength of a wire consists of bending and unbending it until the wire breaks. Considering bending and unbending as two operations, the probability that the wire will break at the rth operation is $(1-p)p^{r-1}$, where $r = 1, 2, 3 \ldots$ and $0 < p < 1$. Find $P(R = n \mid R \geqslant n-1)$.

Solution. $P(R \geqslant n-1) = 1 - \sum_{r=1}^{n-2} (1-p)p^{r-1} = p^{n-2}$. Thus

$$P(R = n \mid R \geqslant n-1) = \frac{(1-p)p^{n-1}}{p^{n-2}} = (1-p)p.$$

Note that this is a discrete analogue of the 'non-ageing' items for which $f(x) = \lambda e^{-\lambda x}, x > 0$. □

Problem 2.7 A device has been designed to measure the strength of subjects. In terms of the units calibrated on the device, the strength S of a given race of male adults can be taken to be a random variable with the probability density function $\lambda e^{-\lambda s}$, $s > 0$, $\lambda > 0$. For a particular task a strong man with a strength of at least s_0 is needed, and men selected at random will be tested one by one till the required strong man is found. Find the probability that exactly n men will be tested.

Solution. The probability that a man tested at random will be the required strong man is $\int_{s_0}^{\infty} \lambda e^{-\lambda s}\, ds = e^{-\lambda s_0}$. For the nth man to be the first strong man, the earlier $(n-1)$ men must have been all 'not strong'. Since the probability that a man tested at random will be not strong is $1-e^{-\lambda s_0}$, the required probability is

$$(1-e^{-\lambda s_0})^{n-1} e^{-\lambda s_0}.$$

Problem 2.8 The time to failure of a system has the probability density function $f(t)$, $t > 0$. The failure can only be detected through inspecting the system at times $x_1, x_2, x_3, \ldots x_x, \ldots$. These inspection times are such that

$$P[\text{system fails in } (x_{i-1}, x_i) \mid \text{system functioned at } x_{i-1}] = p,$$

for $i = 1, 2, 3 \ldots$; $x_0 = 0$. Find the ith inspection time x_i in terms of p. Show that if the failure rate function is monotonic increasing then the intervals between the successive inspection times form a decreasing sequence.

Solution. The condition on the inspection times x_i means

$$\frac{F(x_{i+1}) - F(x_i)}{1-F(x_i)} = p, \; i = 1, 2, 3, \ldots$$

so that $F(x_{i+1}) = p+qF(x_i)$, where $q = 1-p$. Repeated substitution gives

$$F(x_{i+1}) = p+pq+pq^2+\ldots+pq^{i-1}+q^i[p+qF(x_0)].$$
$$= 1-q^{i+1}, \text{ for } x_0 = 0 \text{ gives } F(x_0) = 0.$$

Thus we would find x_i from $x_i = F^{-1}(1-q^i)$.
Let $d_i = x_i - x_{i-1}$ so that $d_i > 0$. We now wish to show that $d_i > d_{i+1}$, $i = 1,2,3\ldots$. Let $S(t) = 1-F(t)$, and let $r(t) = f(t)/S(t)$ be a monotonic increasing function. We now have

$$\int_{x_i-d_i}^{x_i} r(t)\,dt < \int_{x_i}^{x_i+d_i} r(t)\,dt.$$

Using the substitution $u = S(t)$ in integrating $r(t)$, we get

$$\frac{S(x_i+d_i)}{S(x_i)} < \frac{S(x_i)}{S(x_i-d_i)}.$$

Subtraction of one from both sides of this inequality leads to

$$\frac{F(x_i+d_i)-F(x_i)}{1-F(x_i)} > \frac{F(x_i)-F(x_{i-1})}{1-F(x_{i-1})} = p = \frac{F(x_{i+1})-F(x_i)}{1-F(x_i)}.$$

Thus $F(x_i+d_i) > F(x_{i+1})$, so that $x_i+d_i > x_{i+1}$; that is, $d_i > d_{i+1}$. □

***Problem* 2.9** The conditional probability $P(A|x)$ of an event A, given that a random variable X has taken the value x is available. Find $P(A)$.

Solution. Suppose X is discrete and its possible values are x_i, $i = 1,2,3\ldots$ and $P(X = x_i) = P(x_i)$. The events $A \cap x_i$ are mutually exclusive so that

$$P(A) = \sum_i P(A \cap x_i) = \sum_i P(A|x_i)\,P(x_i).$$

If X is continuous with the probability density function $f(x)$, we have

$$P(A) = \int_{-\infty}^{+\infty} P(A|x)\,f(x)\,dx.$$ □

***Problem* 2.10** In an experiment the number of trials, R, that will be carried out is a random variable with $P(R = r) = \theta(1-\theta)^{r-1}, r = 1,2,3\ldots.$ The probability that the experiment will yield a socially useful result, given that r trials were carried out, is $\dfrac{\lambda^r e^{-\lambda}}{r!}$. Find the probability that the experiment will yield the socially useful result.

Solution. Let A be the event that the socially useful result is achieved, so that $P(A|r) = \lambda^r e^{-\lambda}/r!$

$$P(A) = \sum_{1}^{\infty} (\lambda^r e^{-\lambda}/r!)\, \theta(1-\theta)^{r-1}$$
$$= \theta e^{-\lambda}[e^{\lambda(1-\theta)} - 1]/(1-\theta)$$
$$= \theta(e^{-\lambda\theta} - e^{-\lambda})/(1-\theta). \quad \square$$

Problem 2.11 A machine has to be set for the production of a large batch of components. For a given setting, each of the components produced by the machine has a constant probability p of being defective; however the setting process is such that this probability varies from setting to setting and it may be taken to be a random variable with the probability density function $(1-p)^m (m+1)$, $0 < p < 1$. The components are checked by an inspector who makes no errors for good components, but he may pass a defective item as being a good item. If the proportion of defectives in the batch is p then the inspector will pass a defective item as being good with a probability of $kp(1-p)^2$. The machine is now being set for production; what is your forecast of the proportion of defective components in the batch after the inspector has checked the batch?

Solution. The proportion of defectives in the batch will be the proportion of defectives passed as being good by the inspector. Since the batch is large, we take this proportion to be the appropriate probability. Suppose A is the event that the inspector passes a defective item, then $P(A/p) = kp(1-p)^2$, and we would like to compute $P(A)$.

$$P(A) = \int_0^1 P(A|p)\, f(p)\, dp = k(m+1) \int_0^1 p(1-p)^{m+2}\, dp.$$
$$= k(m+1)/(m+3)(m+4). \quad \square$$

2.4 Expectation

2.4.1 A general definition The concept of expectation, or expected value, is a very important one. The term itself shows the influence of gambling on the development of statistics. If $h(X)$ is a function of a random variable X, we denote the expected value of $h(X)$ by $E[h(X)]$ and define it as

$$E[h(X)] = \sum_x h(x)P(x) \quad \text{for discrete } X$$
$$E[h(X)] = \int_{-\infty}^{+\infty} h(x) f(x)\, dx \quad \text{for continuous } X.$$

Problem 2.12 The probability distribution of a random variable R is given in the table. Calculate $E(R^2+2R)/(R+1)$.

r	0	1	2	3	4	5
$P(R = r)$	0·1	0·3	0·1	0·2	0·2	0·1

Solution. $$E\left[\frac{R^2+2R}{R+1}\right] = \sum_{r=0}^{5}\left[\frac{r^2+2r}{r+1}\right]P(r) = 3{\cdot}01.$$ □

Problem 2.13 A coin is tossed until a head appears. Peter will receive £2^r if the first head appears on the rth toss. Find Peter's expected gain. If the rules are altered so that Peter will receive £2^r as long as $r \leqslant 20$ but will receive £2^{20} for $r \geqslant 21$, find Peter's expected gain.

Solution. Let $P(r) = P$ (first head at rth toss). Thus $P(r) = (\frac{1}{2})^r$ and

$$E(\text{gain}) = \sum_{r=1}^{\infty} 2^r(\tfrac{1}{2})^r$$

which is infinite. With the altered rules,

$$E(\text{gain}) = \sum_{1}^{20} 2^r(\tfrac{1}{2})^r + 2^{20}\sum_{21}^{\infty}(\tfrac{1}{2})^r = 21.$$

It is interesting to note that a finite limit, even though it is very large, on Peter's gain, makes Peter's expectation rather small. This classical example is usually called the Petersburg paradox. □

Problem 2.14 A good model for explaining the variation, from item to item, of a quality characteristic of a certain manufactured product is a random variable X with p.d.f. proportional to x, in the range $0 < x < \lambda$, and 0 elsewhere. Each of the manufactured items is tested and items for which $X > l$, where $0 < l < \lambda$, are passed and the rest are rejected. The cost due to a rejected item is £C_1 = £$(a\lambda+b)$ and the profit due to a passed item is £(C_2-C_1). The parameter λ is a manufacturing process variable and can be adjusted to any desired value. Find λ such that the expected profit is maximized.

Solution. Let $f(x) = Kx$, $0 < x < \lambda$. $\int_0^{\lambda} Kx\,dx = 1$ gives $K = 2/\lambda^2$. Now the expected profit S is given by

$$\begin{aligned} S &= (C_2-a\lambda-b)\int_l^{\lambda} 2x/\lambda^2\,dx-(a\lambda+b)\int_0^l 2x/\lambda^2\,dx \\ &= C_2(1-l^2/\lambda^2)-a\lambda-b. \end{aligned}$$

$dS/d\lambda = 0$ gives $\lambda = (a/2C_2l^2)^{-\frac{1}{3}}$. Note that $d^2S/d\lambda^2 = -6C_2l^2/\lambda^4 < 0$, so that a maximum has been found. □

2.4.2 Some particular expected values: moments One particularly useful form is the expected value of a power of X. A common term for describing $E[X^k]$ is that it is the kth moment about the origin; $E[(X-c)^k]$ would be described as the kth moment about c. We note that:

1. $E[(X-c)^k] = \sum_x (x-c)^k P(x)$ for discrete X
2. $E[(X-c)^k] = \int_{-\infty}^{+\infty} (x-c)^k f(x)\,dx$ for continuous X.
3. $E(X)$ is often called the mean of X and it is convenient to denote it by μ.
4. It is convenient to denote $E[(X^k)]$ by μ'_k. Note that $\mu'_1 = \mu$.
5. A useful value for c is μ, and $E[(X-\mu)^k]$ is often called the kth moment about the mean. It is convenient to denote this moment by μ_k. Note that $\mu_1 = 0$.

Moments, when they exist, provide a useful summary of the nature of a random variable. Sometimes it is possible to find the moments but not the probability distribution or the probability density function explicitly. In practice, we rarely use a value of k beyond 4. Looking at $E(X)$, we see that it is a weighted average of x, and as such it is an indication of the order of magnitude of X. $E[(X-\mu)^2]$ is a weighted average of the square of the distance of x from μ, and as such it is an indication of the extent to which the values of X are spread out from μ. This measure of spread, called the variance of X, has turned out to be rather more useful than the perhaps more obvious measure $E[|X-\mu|]$. Variance of X is commonly denoted by $V(X)$ or σ^2. Since units of $V(X)$ are the square of the units of X, it is useful to work with the positive square root of $V(X)$, called the standard deviation of X. We denote the standard deviation of X by σ and note that $\sigma \geqslant 0$. In practice we often compute $V(X)$ from

$$V(X) = E[(X^2)] - [E(X)]^2 \quad \text{or} \quad \sigma^2 = \mu'_2 - (\mu'_1)^2$$

If a probability distribution, or a probability density function is symmetrical, moments of odd order (assuming they exist) would be 0 and thus the third moment can be used as a measure of a symmetry. It is usual to use μ_3/σ^3 as a measure of the degree of a symmetry, and this measure is called the coefficient of skewness. Note that this coefficient of skewness does not depend on the unit in which X is measured. Usually we denote this coefficient of skewness by $\sqrt{\beta_1(X)}$. The measure $\beta_2(X) = \mu_4/\sigma^4$ is used as a measure of 'flatness' and it is called the coefficient of kurtosis.

The use of the first two moments leads to an interesting probability statement, usually called Chebyshev's inequality, namely

$$P(|X-\mu| \geqslant t\sigma) \leqslant 1/t^2$$

Thus $1/t^2$ is an upper bound for $P(|X-\mu| \geqslant t\sigma)$.

Problem 2.15 A random variable R takes the value r with probability ar, $r = 1, 2, \ldots, l$. Find $E(R)$ and $V(R)$.

Solution. $\sum_1^l ar = a\, l(l+1)/2 = 1$ gives $a = 2/l(l+1)$. Hence

$$E(R) = \sum_1^l ar^2 = \tfrac{1}{6}al(l+1)(2l+1) = (2l+1)/3$$

$$E(R^2) = \sum_1^l ar^3 = a\, l^2(l+1)^2/4 = l(l+1)/2$$

so that

$$V(R) = l(l+1)/2-(2l+1)^2/9 = \tfrac{1}{18}(l^2+l-2). \qquad \square$$

Problem 2.16 A random variable X has the probability density function $f(x) = kx^\theta$, $0 < x < \theta$. Find $E(X)$ and $V(X)$.

Solution. Now $\int_0^\theta kx^\theta\, dx = k\theta^{+1}/(\theta+1) = 1$ gives $k = (\theta+1)/\theta^{\theta+1}$. Hence

$$E(X) = \int_0^\theta kx^{\theta+1}\, dx = \theta(\theta+1)/(\theta+2)$$

$$E(X^2) = \int_0^\theta kx^{\theta+2}\, dx = (\theta+1)\theta^2/(\theta+3)$$

so that

$$V(X) = E(X^2)-[(E(X)]^2 = \theta^2(\theta+1)/(\theta+2)^2(\theta+3). \qquad \square$$

Problem 2.17 A random variable X has p.d.f. $f(x) = 1/2a$, $-a < x < a$. Find μ_r and hence find the coefficients of skewness and kurtosis.

Solution. $\mu = E(X) = \int_{-a}^{a} x/2a\, dx = 0$. Thus $\mu_r = E(X^r) = |x^{r+1}/2a(r+1)|_{-a}^{a}$ so that $\mu_r = 0$ if r is odd. This is a reflection of the symmetry of $f(x)$. The coefficient of skewness is zero. Now $\mu_{2r} = a^{2r}/2r+1$, $r = 1, 2, 3 \ldots$. Hence the coefficient of kurtosis β_2 is: $\beta_2 = \mu_4/\mu_2^2 = \frac{9}{5}$. $\square$

Problem 2.18 A random variable X has p.d.f. $f(x) = 1/2a$, where $-a < X < a$. Find the probability that $|X| \geqslant 1{\cdot}5\sqrt{V(X)}$ and compare this with the Chebyshev's upper bound.

Solution. We have $E(X) = 0$ and $V(X) = a^2/3$. Now

$$\begin{aligned} P(|X| > 1{\cdot}5a/\sqrt{3}) &= P(|X| > 0{\cdot}866a) \\ &= 2P(X > 0{\cdot}866a) \qquad \text{from symmetry} \\ &= 0{\cdot}134. \end{aligned}$$

Chebyshev's inequality gives

$$P(|X| \geqslant 1{\cdot}5a/\sqrt{3}) \leqslant (1/1{\cdot}5)^2 = 0{\cdot}444.$$

Thus this upper bound is rather wide in relation to the exact probability computed from a full knowledge of the density function. □

2.4.3 Generating functions The expected value of z^X, where z is a real variable, is a useful one, especially for integral-valued random variables. This expected value is called the probability generating function (p.g.f.), and we shall denote it by $G(z)$. We take the case of an integral-valued random variable R to demonstrate some of the properties of $G(z)$.

1. $G(z) = E[(z^R)] = \sum_{r=0}^{\infty} z^r P(r)$.
2. $G(1) = 1$.
3. $G'(z) = \sum rz^{r-1} P(r)$, so that $E(R) = G'(1)$.
4. $V(R) = G''(1)+G'(1)-[G'(1)]^2$.

Note that the coefficient of z^r in the expansion of $G(z)$ is $P(r)$.

Another useful expected value, called the moment generating function (m.g.f.), is $M(t) = E(e^{tX})$ where t is a dummy variable. If X is a continuous random variable with p.d.f. $f(x)$, we have

$$M(t) = \int_{-\infty}^{+\infty} e^{tx} f(x)\,dx = 1 + \sum_{k=1}^{\infty} (t^k/k)\,\mu'_k$$

Thus, given $M(t)$, we can find μ'_k, assuming it exists, either as an appropriate coefficient in the expansion of $M(t)$ or by evaluating the kth derivative of $M(t)$ at $t = 0$.

The probability generating function and the moment generating function may or may not exist. However the characteristic function $E(e^{itX})$ always exists and its properties can be easily examined.

A very useful property of the above three generating functions is that to a given generating function there is one corresponding probability distribution (or density function). In many problems a probability distribution is identified by first computing one of these three expected values.

2.4.4 Some further properties

1. $E(a) = a$ if a is a constant.
2. $E(aX) = aE(X)$.
3. $V(aX) = a^2 V(X)$.
4. $E[e^{t(a+X)}] = e^{at}E[(e^{tX})]$, or $M_{X+a}(t) = e^{at}M_X(t)$.
5. $M_{aX}(t) = M_X(at)$.

Problem 2.19 A manufacturing process produces a proportion p of defective items. Items are sampled one by one until exactly 3 defectives are obtained, when sampling terminates. Find the probability distribution of N, the number of non-defective items sampled. Obtain its p.g.f. and hence calculate $E(N)$ and $V(N)$.

Solution. If there are exactly n non-defectives when sampling terminates, then the first $n+2$ items sampled must include n non-defectives and 2 defectives and the $(n+3)$th item must of course, be a defective. Thus with $q = 1-p$, we have

$$P(N = n) = \binom{n+2}{2} q^n p^2 . p = p^3 \binom{n+2}{2} q^n \qquad n = 0, 1, 2, \ldots .$$

The p.g.f. is

$$G(z) = E(z^N) = \sum_{n=0}^{\infty} z^n P(N = n)$$

$$= p^3 \sum_{n=0}^{\infty} \binom{n+2}{2} (qz)^n = \left(\frac{p}{1-qz}\right)^3 .$$

Now $dG/dz = 3p^3 q/(1-qz)^4$, so that

$$E(N) = \left.\frac{dG}{dz}\right|_{z=1} = 3\frac{q}{p}.$$

Also $d^2G/dz^2 = 12p^3q^2/(1-qz)^5$, so that

$$E[N(N-1)] = \left.\frac{d^2G}{dz^2}\right|_{z=1} = \frac{12q^2}{p^2}.$$

Hence

$$V(N) = E[N(N-1)] + E(N) - \{E(N)\}^2$$

$$= \frac{12q^2}{p^2} + \frac{3q}{p} - \frac{9q^2}{p^2} = \frac{3q}{p^2} \qquad \square$$

Problem 2.20 A random variable R takes the value r, $r = 0, 1, 2 \ldots$ and its p.g.f. is $b(1-\alpha z)^{-1}$. Find b, $E(R)$, $V(R)$ and $P(r)$.

Solution. Let $G(z) = b(1-\alpha z)^{-1}$, then $G(1) = 1$ gives $b = 1-\alpha$. Hence

$$E(R) = G'(1) = \alpha/(1-\alpha).$$

$$V(R) = G''(1) + G'(1) - [G'(1)]^2 = \alpha/(1-\alpha)^2.$$

Now $G(z) = (1-\alpha) \sum_r (\alpha z)^r$ so that $P(r) = (1-\alpha)\alpha^r$. $\square$

Problem 2.21 The probability generating function of a random variable R, which takes the values r with probability $P(r)$, for $r = 0, 1, 2, \ldots$ is $G(z)$. Find $H(z)$, given by $H(z) = \sum_0^{\infty} z^r F(r)$, where $F(r) = \sum_{i=0}^{r} P(i)$.

Solution. Now

$$P(r) = \sum_0^r P(i) - \sum_0^{r-1} P(i),$$

$$\sum_r z^r P(r) = \sum_r z^r \sum_0^r P(i) - z \sum_r z^{r-1} \sum_0^{r-1} P(i)$$

That is, $G(z) = H(z) - zH(z)$ or $H(z) = G(z)/(1-z)$.

Note that the coefficient of z^r in the expansion of $H(z)$ is $P(R \leqslant r)$, so that $H(z)$ will generate the distribution function of R.

***Problem* 2.22** A random variable X has p.d.f. $f(x) = e^{-x}$, $x > 0$. Find its moment generating function and hence μ'_r.

Solution.

$$M(t) = E(e^{tx}) = \int_0^\infty e^{-x(1-t)}\,dx$$

$$= 1/(1-t) = \sum_{r=0}^{\infty} t^r, |t| < 1.$$

Thus μ'_r is the coefficient of $t^r/r!$, i.e. $r!$. □

***Problem* 2.23** A random variable X has the probability density function

$$f(x) = 1+x, \qquad -1 < x < 0$$

$$= 1-x, \qquad 0 < x < 1$$

Find the moment generating function and hence find $E(X)$, $V(X)$.

Solution. Now

$$M(t) = \int_{-1}^{1} e^{tx} f(x)\,dx = \int_{-1}^{0} (1+x)e^{tx}\,dx + \int_0^1 (1-x)e^{tx}\,dx$$

$$= \frac{1}{t^2}(e^t + e^{-t} - 2) = \frac{2}{t^2} \sum_i \frac{t^{2i}}{(2i)!}$$

The coefficient of t is 0, thus $E(X) = 0$. The coefficient of $t^2/2!$ is $\frac{1}{6}$, thus $V(X) = E(X^2) = \frac{1}{6}$.

2.5 Truncated Random Variable Suppose a system with n components will fail if one or more of the n components become defective. If R is the number of defective components, we may characterize the system by $P(r)$, $r = 0, 1, 2, \ldots, n$. Now if the system is examined for the number of defective components only when it has failed, we know that $R > 0$ whenever an observation on R is to be made. Since $P(R > 0) = 1 - P(0)$, we may characterize this situation by

$$P(r \mid R > 0) = P(r)/(1-P(0)), \qquad r = 1, 2, \ldots, n.$$

Note that:

1. $0 \leqslant P(r \mid R > 0) \leqslant 1, \quad r = 1, 2, \ldots, n.$

2. $\sum_{r=1}^{n} P(r \mid R > 0) = 1.$

Thus $\{r, P(r \mid R > 0)\}, r = 1, 2, \ldots, n$ is a proper probability distribution and we may consider the various functions defined for a discrete random variable in sections 2.2 and 2.4. For example,

$$E(R \mid R > 0) = \sum_{r=1}^{n} r\, P(r \mid R > 0).$$

This is an example of an experiment in which a random variable is *truncated.* In this particular case, we have had a discrete random variable truncated to the left at 0. Truncation of this R to the right at $b \leqslant n$ would lead to the characterization

$$P(r \mid R < b) = P(r)/F(b-1), \qquad r = 0, 1, 2, \ldots, (b-1).$$

Truncation to the left and to the right of a discrete random variable X would lead to the characterization

$$P(x \mid a < X < b) = P(x)/P(a < X < b), \qquad a < x < b.$$

If $f(x)$ is the p.d.f. of a continuous random variable X, we may write the p.d.f. of X truncated to the left at a as

$$g(x) = 0, \qquad x \leqslant a$$
$$= \frac{f(x)}{(1-F(a))}, \; x > a.$$

In the case of truncation to the left and to the right, we would have

$$g(x) = 0, \qquad x \leqslant a, x \geqslant b,$$
$$= f(x)/[F(b)-F(a)], \; a < x < b.$$

Note that we may consider the various functions defined for a continuous random variable in sections 2.3 and 2.4 when working with truncated random variables.

Problem 2.24 A random variable R, which takes the value r with probability $k\alpha^r$, $r = 0, 1, 2, \ldots$, is to be truncated to the right at l such that $P(0 \mid R \leqslant l-1)$ is approximately equal to $2P(0)$. Find a suitable value of l.

Solution. Since $\sum_{0}^{\infty} P(r) = \sum_{0}^{\infty} k\alpha^r = 1$, then $k = 1-\alpha$. Now

$$P(R \leqslant l-1) = (1-\alpha)\sum_{0}^{l-1} \alpha^r = 1-\alpha^l.$$

$$P(r\,|\,R \leqslant l-1) = \alpha^r(1-\alpha)/(1-\alpha^l)$$

$$P(0)|R \leqslant l-1) = (1-\alpha)/(1-a^l).$$

Thus $P(0\,|\,R \leqslant l-1) \simeq 2\,P(0)$ means $(1-\alpha)/(1-\alpha^l) \simeq 2(1-\alpha)$ or $l \simeq \ln 0{\cdot}5/\ln \alpha$.

Note that we would round off to get an integral answer. □

Problem 2.25 The 'noise-level' X of a certain type of electronic device, varies from device to device. A reasonable model for this variation is that X is a random variable with p.d.f. $f(x) = 2xe^{-x^2}, x > 0$. If $X > c$, where c is a specified constant, the device is said to be of quality grade II; otherwise the device is said to be of quality grade I. An automatic machine grades these devices as grade I or grade II. A device of grade II is selected at random. Find the probability that its noise-level will exceed $1{\cdot}5c$.

Solution. Let $g(x)$ be the p.d.f. of noise level of grade II devices, so that

$$g(x) = 0, \qquad x \leqslant c$$
$$= 2xe^{-x^2}/(1-F(c)), x > c.$$

Now $F(x) = \int_0^x 2te^{-t^2}\,dt = 1-e^{-x^2}$. This leads to

$$P = P(\text{grade II device has } X > 1{\cdot}5c) = \int_{1\cdot 5c}^{\infty} g(x)\,dx$$
$$= \int_{1\cdot 5c}^{\infty} 2xe^{-x^2}/e^{-c^2}\,dx = e^{-1\cdot 25c^2}.$$ □

Problem 2.26 If for the electronic device of Problem 2.25, the mean and the variance of grade i are given to be μ_i and σ_i^2, $i = 1, 2$, find the mean and the variance of X.

Solution. Now

$$\mu_1 = \frac{1}{1-e^{-c^2}}\int_0^c 2x^2e^{-x^2}\,dx.$$

This gives $\int_0^c 2x^2c^{-x^2}\,dx = (1-e^{-c^2})\mu_1$. Similarly $\int_c^{\infty} 2x^2e^{-x^2}\,dx = e^{-c^2}\mu_2$, so that

$$\mu = \int_0^{\infty} 2x^2e^{-x^2}\,dx = \mu_1 + e^{-c^2}(\mu_2-\mu_1).$$

For computing the variance of X, we exploit the result $V(X) = E(X^2) - [E(X)]^2$. For grade I devices we have

$$E(X^2\,|\,X < c) = \sigma_1^2+\mu_1^2 = \frac{1}{1-e^{-c^2}}\int_0^c 2x^3e^{-x^2}\,dx$$

Therefore

$$\int_0^c 2x^3e^{-x^2}\,dx = (1-e^{-c^2})(\sigma_1^2+\mu_1^2).$$

Similarly

$$\int_c^\infty 2x^3e^{-x^2}\,dx = e^{-c^2}(\sigma_2^2+\mu_2^2).$$

Thus

$$E(X^2) = \int_0^\infty 2x^3e^{-x^2}\,dx = \int_0^c 2x^3e^{-x^2}\,dx+\int_c^\infty 2x^3e^{-x^2}\,dx$$

$$= \sigma_1^2+\mu_1^2+e^{-c^2}(\sigma_2^2-\sigma_1^2+\mu_2^2-\mu_1^2).$$

$$V(X)=E(X^2)-\mu^2 = E(X^2)-[\mu_1^2+2\mu_1 e^{-c^2}(\mu_2-\mu_1)+e^{-2c^2}(\mu_2-\mu_1)^2]$$

$$= \sigma_1^2+e^{-c^2}[\sigma_2^2-\sigma_1^2+(1-e^{-c^2})(\mu_2-\mu_1)^2].$$ □

Note: Suppose in Problem 2.25, the noise level X has an arbitrary distribution, and the proportions of grade I and grade II quality devices are p and q respectively ($p+q = 1$). If the means and the variances of the noise level of the two grades are μ_1, σ_1^2 and μ_2, σ_2^2, the mean and variance of X are given by

$$\mu = p\mu_1+q\mu_2,\ \sigma^2 = p\sigma_1^2+q\sigma_2^2+pq(\mu_1-\mu_2)^2.$$

EXERCISES

1. A random variable X has a p.d.f. $f(x) = 2x/a^2$, $0 < x < a$. Another random variable R takes the value r, $r = 0, 1, 2, \ldots, 9$, with probability $P(r) = P[0{\cdot}1ar < X < 0{\cdot}1a(r+1)]$. Find (i) $P(r)$, (ii) $E(R)$, (iii) $V(X)$.

2. In a qualifying round of a competition for places on a national darts team, each of the contestants is allowed to throw up to 3 darts. Two darts within a distance of 60 millimeters from the bulls-eye is the standard for passing the qualifying round. Peter's skill is such that for each of his throws, X, the distance in millimeters from the bull's-eye, is a random variable with probability density function $f(x) = kx^2e^{-x^3}$, $0 < x < 100$. Find, (i) k, (ii) P (Peter qualifies in 3 throws), (iii) P (Peter does not qualify).

3. The voltage across an electrical circuit is a random variable with p.d.f. $f(v) = e^{-(v-k)}$, $v > k > 0$. The resistance of the circuit is fixed at a known value of Ω ohms. Assuming ohm's law, which says that current = voltage/resistance, find the mean and the variance of the current through the circuit.

4. For the random variable X of exercise 1, find the Chebyshev's bound

for the probability $P(|X-E(X)| \geqslant a/3)$, and compare it with the exact value of P.

5. For the random variable R of exercise 1, find $P(R = 8 | R > 7)$.

6. For the random variable X of exercise 1, find $E(e^{tx} | X < a/2)$ and hence show that $E(X | X < a/2) = a/3$.

Chapter 3

Random Variable: Multivariate Case

3.1 Definition Frequently two or more characteristics are observed in an experiment. For example, an engineer may compute the running temperature and the efficiency of a machine; a doctor may measure the temperature, the blood pressure, and the pulse rate of a patient. Each of the possible outcomes of such an experiment is usually described by two or more real numbers. Generally, then, we consider an experiment E and its associated sample space S. Let $X_1, X_2, X_3, \ldots, X_k$ be k functions such that each assigns a real number to every element s of S. We call $(X_1, X_2, \ldots, X_n)$ a k-dimensional random variable. (Another name is a k-dimensional random vector).

It is convenient to consider a two-dimensional random variable. Most of the ideas for this special case of $k = 2$ can be easily adapted for the case of $k > 2$.

3.2 Discrete Random Variable We say that (X, Y) is a discrete random variable if the possible values of (X, Y) are finite or countably infinite. Thus in a discrete case the possible values of (X, Y) may be listed as (x_i, y_j), $i = 1, 2, 3, \ldots; j = 1, 2, 3, \ldots$. This list would terminate in the finite case. We associate a real number $P(x, y)$ with each of the possible values (x, y), and call it the probability that $X = x$, $Y = y$. We note that

1 $0 \leqslant P(x_i, y_j) \leqslant$ for all i, j.

2. $\sum_{i=1}^{\infty} \sum_{j=1}^{\infty} P(x_i, y_j) = 1.$

The collection of triplets $[x, y, P(x, y)]$ is called the probability distribution of (X, Y). The cumulative distribution function (c.d.f.) of a random variable (X, Y) is defined by

$$F(x, y) = \sum_i \sum_j P(x_i, y_j),$$

where the summation is over the values of i, j such that $x_i \leqslant x$, $y_j \leqslant y$.

Given a random variable (X, Y), we may wish to consider X and Y individually. Probability statements about X alone can be made from the *marginal probability distribution of* X, defined by

$$P(x_i) = \sum_{j=1}^{\infty} P(x_i, y_j), \qquad i = 1, 2, 3, \ldots.$$

The marginal distribution of Y is given by

$$P(y_j) = \sum_{i=1}^{\infty} P(x_i, y_j), \qquad j = 1,2,3,\ldots.$$

Often we are interested in X when it is known that Y has taken a particular value y_j; the *conditional probability distribution* of X given that $Y = y_j$, defined below, is useful in such cases.

$$P(x_i \mid y_j) = P(x_i, y_j)/P(y_j), \qquad i = 1,2,3,\ldots.$$

Note that

1. $0 \leqslant P(x_i \mid y_j) \leqslant 1$ for all i (and j).

2. $\sum_i P(x_i \mid y_j) = [\sum_i P(x_i, y_j)]/P(y_j) = 1.$

The conditional probability distribution of Y, given that X has taken a particular value x_i is given by

$$P(y_j \mid x_i) = P(x_i, y_j)/P(x_i), \qquad j = 1,2,3,\ldots.$$

Independence of random variables X, Y is an important idea. Informally speaking, X, Y would be considered to be independent if the probability statements about X are not influenced by the knowledge that Y has taken a value y_j. Formally we say that X, Y are independent if and only if

$$P(x_i, y_j) = P(x_i)P(y_j), \qquad \text{for all } i, j.$$

Problem 3.1 The probability distribution of a random variable (X, Y) is given by $P(x_i, y_j) = C(i+j)$, $i = 1,2,3,4$; $j = 1,2,3$. Find (i) C, (ii) The marginal probability distributions of X, Y, (iii) The conditional probability distribution of X, given $Y = y_j$ and the conditional probability distribution of Y, given $X = x_i$. Show that X, Y are not independent.

Solution. (i) $C \sum_{i=1}^{4} \sum_{j=1}^{3} (i+j) = C(\sum_i \sum_j i + \sum_i \sum_j j) = C(30+24) = 54C.$

Therefore $\sum_i \sum_j P(x_i, y_j) =$ gives $C = \frac{1}{54}$.

(ii) $$P(x_i) = \sum_j P(x_i, y_j) = \tfrac{1}{54} \sum_j (i+j) = (i+2)/18.$$

$$P(y_j) = \sum_i P(x_i, y_j) = \tfrac{1}{54} \sum_i (i+j) = (2j+5)/27.$$

(iii) $$P(x_i \mid y_j) = P(x_i, y_j)/P(y_j) = (i+j)/(4j+10).$$

$$P(y_j \mid x_i) = P(x_i, y_j)/P(x_i) = (i+j)/(3i+6).$$

Note that $\sum_{i=1}^{4} P(x_i \mid y_j) = 1$ and $\sum_{j=1}^{3} P(y_j \mid x_i) = 1.$

Now $P(x_i)P(y_j) = (i+2)(10+4j)/(18)(54) \neq P(x_i, y_j)$. Therefore X, Y are not independent. □

Problem 3.2 Two dice are thrown. The random variable X represents the sum of the two scores and the random variable Y represents the number of sixes. Find (i) The probability distribution of (X, Y), (ii) The marginal probability distributions of X and Y, (iii) The probability that X will exceed 7, given that $Y = 1$.

Solution. (i) The possible values for X are $2, 3, 4, \ldots, 12$ and the possible values for Y are $0, 1, 2$. If we think of the two dice being labelled 1, 2 and the scores on them as R_1, R_2, we see that the event $X = 2$, $Y = 0$ will occur when $R_1 = 1$ and $R_2 = 1$. Now $P(R_1 = 1) = \frac{1}{6}$ and $P(R_2 = 1) = \frac{1}{6}$; also it is reasonable to suppose that R_1, R_2 are independent. Thus $P(X = 2, Y = 0) = P(R_1 = 1, R_2 = 1) = \frac{1}{36}$. Similarly we can compute the remaining probabilities $P(x, y)$. The complete probability distribution of (X, Y) is shown in Table 3.1.

Table 3.1 Values of $P(x, y)$.

y \ x	2	3	4	5	6	7	8	9	10	11	12	Total
0	$\frac{1}{36}$	$\frac{2}{36}$	$\frac{3}{36}$	$\frac{4}{36}$	$\frac{5}{36}$	$\frac{4}{36}$	$\frac{3}{36}$	$\frac{2}{36}$	$\frac{1}{36}$	0	0	$\frac{25}{36}$
1	0	0	0	0	0	$\frac{2}{36}$	$\frac{2}{36}$	$\frac{2}{36}$	$\frac{2}{36}$	$\frac{2}{36}$	0	$\frac{10}{36}$
2	0	0	0	0	0	0	0	0	0	0	$\frac{1}{36}$	$\frac{1}{36}$
Total	$\frac{1}{36}$	$\frac{2}{36}$	$\frac{3}{36}$	$\frac{4}{36}$	$\frac{5}{36}$	$\frac{6}{36}$	$\frac{5}{36}$	$\frac{4}{36}$	$\frac{3}{36}$	$\frac{2}{36}$	$\frac{1}{36}$	1

(ii) The column totals displayed in the last row of Table 3.1 form the marginal probability distribution of X and the row totals displayed in the last column form the marginal probability distribution of Y.

(iii) $P(X > 7 \mid Y = 1) = \sum_{i=8}^{12} [P(i, 1)/P(Y = 1)] = (\frac{8}{36})/(\frac{10}{36}) = 0{\cdot}8$. □

Problem 3.3 In a clinical trial the response of the patients to a given treatment is classified as good $(= 1)$, fair $(= 2)$, and poor $(= 3)$. For a given class of patient, the probability of response i is p_i, $i = 1, 2, 3$; $\sum_{i=1}^{3} p_i = 1$. The treatment is given to n patients from this class. If R_i, $i = 1, 2, 3$, is the number of patients out of n who show the response i, find

(i) the probability distribution of (R_1, R_2, R_3),
(ii) $P(R_1 = r_1)$,
(iii) the probability that the number of patients showing response 1 will exceed the number of patients showing response 2 by exactly $k > 0$.

Solution. (i) It is reasonable to suppose that each patient has independently a probability p_i of response i. The probability of exactly r_i with response i, in a particular order (for example, r_1 successive responses 1, followed by r_2 responses 2, followed by r_3 responses 3), is $p_1^{r_1} p_2^{r_2} p_3^{r_3}$, $\sum_{i=1}^{3} r_i = n$. There are $n!/(r_1!\,r_2!\,r_3!)$ ways of observing r_1, r_2, r_3 so that the required probability distribution is given by

$$P(r_1, r_2, r_3) = \frac{n!}{r_1!\,r_2!\,r_3!} p_1^{r_1} p_2^{r_2} p_3^{r_3}, \quad 0 \leqslant r_i \leqslant n, \ \sum_{i=1}^{3} r_i = n$$

(ii) Now $r_3 = n - r_1 - r_2$ and $p_3 = 1 - p_1 - p_2$. Thus we may write $P(r_1, r_2, r_3)$ as

$$P(r_1, r_2) = \frac{n!}{r_1!\,r_2!\,(n - r_1 - r_2)!} p_1^{r_1} p_2^{r_2} (1 - p_1 - p_2)^{n - r_1 - r_2}$$

$$P(r_1) = \sum_{r_2 = 0}^{n - r_1} P(r_1, r_2) = \frac{n!}{r!(n - r_1)!} p_1^{r_1} (1 - p_1)^{n - r_1}.$$

Note that we could have obtained $P(r_1)$ by arguing that there are two possible responses for a patient, '1' and 'not 1', with probability p_1 and $1 - p_1$ respectively.

(iii) Let $q = P(R_1 - R_2 = k > 0)$. Now

$$q = \sum_{r} P(R_1 = r, \quad R_2 = r - k, \quad R_3 = n - 2r + k) \qquad \square$$

Problem 3.4 The probability distribution of a random variable (X, Y, Z) is given by $P(x_i, y_j, z_k) = C(i + j + k)$, $i = 1, 2, 3, 4$; $j = 1, 2, 3$; $k = 1, 2$. Find (i) C, (ii) The probability distribution of (X, Y), (iii) The marginal probability distribution of Z, (iv) The conditional probability distribution of (X, Y), given that $Z = z_k$, (v) The conditional probability distribution of Z given that $X = x_i$, $Y = y_j$.

Solution. (i) $C \sum_{i=1}^{4} \sum_{j=1}^{3} \sum_{k=1}^{2} (i + j + k) = 144C$, so that $C = 1/144$.

(ii) $P(x_i, y_j) = (1/144) \sum_{k} (i + j + k) = (1/144)(2i + 2j + 3)$

$= (1/24)[\{(i+j)/3\}+\tfrac{1}{2}]$.

(iii) $P(z_k) = (1/144)\sum_i\sum_j (i+j+k) = (1/144)(54+12k) = (k/12)+\tfrac{3}{8}$.

(iv) $P(x_i, y_j | z_k) = P(x_i, y_j, z_k)/P(z_k) = (i+j+k)/(54+12k)$.

(v) $P(z_k | x_i, y_j) = P(x_i, y_j, z_k)/P(x_i, y_j) = (i+j+k)/(2i+2j+3)$. □

3.3 Continuous Random Variable We say that a random variable (X, Y) is a continuous random variable if (X, Y) can have any value in a given non-countable set of the Euclidean plane. An example of this non-countable set is $-\infty < x < \infty$, $-\infty < y < \infty$; another example is the rectangle $a \leqslant x \leqslant b$, $c \leqslant y \leqslant d$. The function $P(x, y)$, which means $P(X = x, Y = y)$ ceases to be a useful one and we work with the cumulative distribution function F, and the probability density function f. These two functions are defined by $F(x, y) = P(X \leqslant x, Y \leqslant y)$, $f(x, y) = \partial F(x, y)/\partial x \partial y$, assuming that F is differentiable.
Note that

1. $f(x, y) \geqslant 0$.
2. $F(x, y) = \int_{-\infty}^{y}\int_{-\infty}^{x} f(x, y)\,dxdy$.
3. $\int_{-\infty}^{+\infty}\int_{-\infty}^{+\infty} f(x, y)\,dxdy = 1$.

Given a random variable (X, Y) we may wish to consider X and Y individually. We define the marginal probability density functions of X and Y as

$$u(x) = \int_{-\infty}^{+\infty} f(x, y)\,dy, \qquad v(y) = \int_{-\infty}^{+\infty} f(x, y)\,dx.$$

Note, for example that $P(a < x < b) = \int_a^b u(x)\,dx$. Often we are interested in making probability statements about X when it is known that Y has taken a particular value y; the conditional probability density function, $u(x | y)$, defined below, is useful in such cases.

$$u(x | y) = f(x, y)/v(y), \qquad v(y) > 0.$$

Similarly, $$v(y | x) = f(x, y)/u(x), \qquad u(x) > 0$$

Note that these conditional densities satisfy all the requirements for a univariate probability density function. For example, $u(x|y) \geqslant 0$ $\int_{-\infty}^{+\infty} u(x | y)\,dx = 1$.

In the case of continuous random variable (X, Y) the important idea of independence is formalized by the statement that X, Y are said to be independent if and only if $f(x, y) = u(x)v(y)$.

Problem 3.5 The joint p.d.f. of a continuous random variable (X, Y) is

$f(x, y) = C(x+y)$, $0 < x < a$, $0 < y < b$. Find (i) C, (ii) the marginal probability densities, (iii) the conditional probability densities, (iv) $P(X > a/2 \mid Y > b/2)$.

Are X, Y independent?

Solution. (i) $C \int_0^b \int_0^a (x+y)\, dxdy = 1$ gives $C = 2/[ab(a+b)]$.

(ii) $u(x) = C \int_0^b (x+y)\, dy = bC(x+b/2) = 2(x+b/2)/a(a+b)$.

$v(y) = C \int_0^a (x+Y)\, dx = 2(y+a/2)/b(a+b)$.

(iii) $u(x \mid y) = f(x, y)/v(y) = (x+y)/(y+a/2)$.

$v(y \mid x) = f(x, y)/u(x) = (x+y)/(x+a/2)$.

(iv) $P(X > a/2 > b/2) = P(X > a/2, Y > b/2)/P(Y > b/2)$

$= 3(a+b)/4(a+3b/2)$.

X, Y are not independent since $f(x, y) \neq u(x)\, v(y)$. □

Problem 3.6 The lengths X_1 and X_2 of the sides of a rectangle are independent random variables and each has the probability density function $f_i(x_i) = 1$; $0 < x_i < 1$, $i = 1, 2$. Find the probability that the area of the rectangle will exceed $\frac{1}{2}$.

Solution. The area will exceed $\frac{1}{2}$ if the values, X_1 and X_2 taken by the random variables are such that $x_1 x_2 > \frac{1}{2}$ where $0 < x_1, x_2 < 1$.

Now X_1, X_2 are independent, so that the joint p.d.f. is given by

$$f(x_1, x_2) = f_1(x_1) f_2(x_2) = 1,$$

Thus the required probability P is

$$P = \int_{x_2 = \frac{1}{2}}^{1} \int_{x_1 = 1/2x_2}^{1} dx_1\, dx_2 = \tfrac{1}{2}(1 - \ln 2)$$ □

Problem 3.7 The time to failure, X, of a mechanical component has a probability distribution $f(x) = \alpha x^{\alpha-1} \exp(-x^\alpha)$, where $\alpha > 1$ and $x > 0$. The time to failure Y of an electrical component has a probability distribution $g(y) = \beta \exp(-\beta y)$, where $\beta > 0$, $y > 0$. These two components together make up a system which will fail as soon as one of the two components fail. Assuming X, Y to be independent, find the probability that the system will fail in the interval $(0, t)$.

Solution. We note that the system will survive t if $X > t$ and $Y > t$. Thus the required probability is

$$P = 1 - P(X > t, Y > t) = 1 - \int_t^{\infty} \int_t^{\infty} f(x,y)\, dxdy.$$

The joint density $f(x, y) = f(x) g(y)$ since X, Y are independent. Thus

$$P = 1 - \int_t^{\infty} \int_t^{\infty} \alpha\beta \exp(-\beta y) x^{\alpha-1} \exp(-x^\alpha)\, dxdy = 1 - \exp[-(t^\alpha + \beta t)]$$ □

Problem 3.8 A random variable (X, Y) represents (Peter's gain, Paul's gain) in a gamble, and it has the p.d.f. $f(x, y) = (x+y+2)/8, -1 < x < +1; -1 < y < +1$. Find the probability that Paul's gain is larger than Peter's gain.

Solution. We need

$$P(X < Y) = \tfrac{1}{8}\int_{-1}^{1}\int_{x}^{1}(x+y+2)\,dydx$$
$$= \tfrac{1}{8}\int_{-1}^{1}(2{\cdot}5-x-1{\cdot}5x^2)\,dx = \tfrac{1}{2}.$$

□

3.4 Expectation

3.4.1 A general defintion If $h(X, Y)$ is a function of a random variable (X, Y), we denote the expected value of $h(X, Y)$ by $E[h(X, Y)]$, and define it as:

$$E[h(X, Y)] = \sum_{i=1}^{\infty}\sum_{j=1}^{\infty} h(x_i, y_j)P(x_i, y_j) \qquad \text{for discrete } (X, Y)$$

$$E[h(X, Y)] = \int_{-\infty}^{+\infty}\int_{-\infty}^{+\infty} h(x, y)f(x, y)\,dxdy \qquad \text{for continuous } (X, Y)$$

Problem 3.9 Show that for a discrete random variable (X, Y), $E[(X \pm Y)] = E(X) \pm E(Y)$, where $E(X)$ and $E(Y)$ are the expected values from the marginal distributions.

Solution.

$$E[(X \pm Y)] = \sum_i\sum_j (x_i \pm y_j)P(x_i, y_j)$$
$$= \sum_i\sum_j x_i P(x_i, y_j) \pm \sum_i\sum_j y_j P(x_i, y_j)$$
$$= \sum_i x_i \sum_j P(x_i, y_j) \pm \sum_i y_j \sum_j P(x_i, y_j)$$
$$= \sum_i x_i P(x_i) \pm \sum_j y_j P(y_j) = E(X) \pm E(Y).$$

Similarly we can show that $E[(X \pm Y)] = E(X) \pm E(Y)$ for continuous (X, Y).

□

Problem 3.10 The random variable (X, Y) of Problem 3.1 with $x_i = i$, $y_j = j$, represents (Peter's, Paul's) score in a game. The agreed stakes are that Peter will get £(Peter's score – Paul's score). Find Peter's expected gain.

Solution.

$$E(\text{Peter's gain}) = E[(X-Y)] = \sum_{i=1}^{4}\sum_{j=1}^{3}(i-j)(i+j)/54 = \sum_i\sum_j (i^2-j^2)/54$$
$$= (1/54)\sum_i (3i^2 - 14) = £17/27$$

Note that from the marginal distributions we have

$$E(X) = \sum_{i=1}^{4} i(i+2)/18 = 25/9 \qquad E(Y) = \sum_{j=1}^{3} j(2j+5)/27 = 58/27$$

Therefore $E(X-Y) = E(X)-E(Y)$ (from Problem 3.9) $= 17/27$ as before □

Problem 3.11 Show that for a discrete random variable (X, Y), where X, Y are independent, $E(XY) = E(X)E(Y)$.

$$\begin{aligned} E(XY) &= \sum_i \sum_j x_i y_j P(x_i, y_j) \\ &= \sum_j \sum_j x_i y_j P(x_i)P(y_j), \qquad \text{since } X, Y \text{ are independent} \\ &= \sum_i x_i P(x_i) \sum_j y_j P(y_j) = E(X)E(Y). \end{aligned}$$

Similarly we can show that $E(XY) = E(X)E(Y)$ for a continuous (X, Y) if X, Y are independent. □

3.4.2 Some particular expected values: moments As in the case of a random variable X, expected values of integral powers of X, Y are particularly useful for a random variable (X, Y). A common term for describing $E[X^l Y^m]$ is that it is the (l, m)th moment about the origins. A convenient notation is

$$\begin{aligned} \mu'_{l,m} &= E(X^l Y^m) = \sum_i \sum_j x_i^l y_j^m P(x_i, y_j) && \text{for discrete } (X, Y) \\ &= \int_{-\infty}^{+\infty} \int_{-\infty}^{+\infty} x^l y^m f(x, y)\, dx dy && \text{for continuous } (X, Y) \end{aligned}$$

Moments about the means, $E(X) = \mu_X$, $E(Y) = \mu_Y$, are denoted by

$$\begin{aligned} \mu_{l,m} &= E[\{(X-\mu_X)^l\}\{Y-\mu_Y\}^m] \\ &= \sum_i \sum_j (x_i-\mu_X)^l (y_j-\mu_Y)^m P(x_i, y_j) && \text{for discrete } (X, Y) \\ &= \int_{-\infty}^{+\infty} \int_{-\infty}^{+\infty} (x-\mu_X)^l (y-\mu_Y)^m f(x, y)\, dx dy && \text{for continuous } (X, Y) \end{aligned}$$

μ_{11}, called the *covariance* of (X, Y), is a particularly useful moment. It may be convenient to compute μ_{11} from

$$\mu_{11} = E[\{X-\mu_X\}\{Y-\mu_Y\}] = E(XY)-\mu_X \mu_Y.$$

It is convenient to use the notation $C(X, Y) = \mu_{11}$.

Correlation coefficient ρ defined by

$$\rho = C(X, Y)/\surd(V(X)V(Y)),$$

is a useful measure of a linear association between X and Y. It can be shown that $-1 \leqslant \rho \leqslant +1$. As an example of a linear association, consider a discrete random variable X with probability distribution $\{x_i, P(x_i)\}$,

$i = 1, 2, 3, \ldots$. Let $y_i = Ax_i + B$, where A, B are constants. We have thus generated a discrete random variable $Y = AX + B$. In general, if a random variable (X, Y) is such that there is a perfect linear relationship between X and Y, so that the values (x_i, y_i) will lie on a straight line, we would have $\rho^2 = 1$. Values of ρ^2 close to 1 indicate a high degree of linearity; values of ρ^2 close to 0 indicate a low degree of linearity. Positive values of ρ mean that Y tends to increase as X increases, and negative values of ρ mean that Y tends to decrease as X increases. If $\rho = 0$, we say that X, Y are uncorrelated.

In the case of independent variables X, Y we have, from Problem 3.11, $E(XY) = E(X)E(Y)$. Thus $C(X, Y) = 0$ and $\rho = 0$. In other words, if X, Y are independent, then X, Y are uncorrelated. However, in general the converse is not true; we could have $\rho = 0$ and X, Y not independent.

Problem 3.12 Random variables X, Y are such that $Y = AX + B$, where A and B are constants. Show that $\rho^2 = 1$.

Solution. Now
$$\begin{aligned} C(X, Y) &= E(XY) - E(X)\,E(Y) \\ &= E[X(AX+B)] - E(X)[AE(X)+B] \\ &= AE(X^2) - A[E(X)]^2 + AV(X). \end{aligned}$$
Also since $Y = AX + B$, $V(Y) = A^2\,V(X)$, so that
$$\rho^2 = \frac{[C(X, Y)]^2}{V(X)\,V(Y)} = \frac{A^2[V(X)]^2}{A^2[V(X)]^2} = 1.$$
Note that $C(X, Y) < 0$ if $A < 0$ so that $\rho = -1$ and $C(X, Y) > 0$ if $A > 0$ so that $\rho = +1$. □

Problem 3.13 Find the correlation coefficient ρ between X and Y for the random variable (X, Y) of problem 3.2.

Solution. The marginal distribution of Y, given on page 33, is $(0, \frac{25}{36})$, $(1, \frac{10}{36})$, $(2, \frac{10}{36})$. Therefore $E(Y) = \frac{1}{3}$, $V(Y) = \frac{10}{36}$. It is instructive to note that Y is the number of sixes in two throws. Now E (number of sixes in one throw) is given by
$$1(\tfrac{1}{6}) + 0(\tfrac{5}{6}) = \tfrac{1}{6}.$$
Hence $E(Y) = \frac{1}{6} + \frac{1}{6} = \frac{1}{3}$ as before. Similarly, $E(X) = 7$, $V(X) = \frac{35}{6}$. Also
$$E(XY) = \tfrac{2}{36}(7+8+9+10+11) + \tfrac{1}{36}(24) = \tfrac{114}{36}$$
Thus $C(X, Y) = E(XY) - E(X)E(Y) = \frac{114}{36} - \frac{7}{3} = \frac{5}{6}$.
so that $\rho = C(X, Y)/\sqrt{(V(X)V(Y))} = \sqrt{\frac{3}{7}}$. □

Problem 3.14 A continuous random variable (X, Y) has p.d.f. $f(x, y) = 1/(ab), 0 < x < a, 0 < y < b$. Find μ'_{lm}. Show that $C(X, Y) = 0$.

Solution.

$$\mu'_{l,m} = E(X^l Y^m) = (1/ab) \int_0^b \int_0^a x^l y^m \, dxdy$$
$$= a^l/[b(l+1)] \int_0^b y^m \, dy = a^l b^m/(l+1)(m+1).$$

The marginal distributions are $u(x) = 1/a$, $0 < x < a$; $v(y) = 1/b$, $0 < y < b$. Since $f(x, y) = u(x)v(y)$, then X, Y are independent. Thus $C(X, Y) = 0$.

Note that we could have computed $E(X^l Y^m)$ from $E(X^l)E(Y^m)$. □

Problem 3.15 The joint p.d.f. of two r.v.s. X and Y is given by

$$f(x, y) = k[(x+y)-(x^2+y^2)], \qquad 0 < x < 1, 0 < y < 1.$$

Show that X and Y are uncorrelated but not independent.

Solution.

$$\int_0^1 \int_0^1 f(x, y) \, dxdy = k \int_0^1 \int_0^1 [(x+y)-(x^2+y^2)] \, dxdy$$
$$= k \int_0^1 [(xy+\tfrac{1}{2}y^2)-(x^2 y+\tfrac{1}{3}y^3)]_0^1 \, dx$$
$$= k[(\tfrac{1}{2}x^2+\tfrac{1}{2}x)-(\tfrac{1}{3}x^3+\tfrac{1}{3}x)]_0^1 = \tfrac{1}{3}k = 1.$$

Thus $k = 3$. Now the marginal p.d.f. of X, say $u(x)$, is

$$u(x) = \int_0^1 f(x, y) \, dy = 3(\tfrac{1}{6}+x-x^2)$$

and the marginal p.d.f. of Y, say $v(y)$ is

$$v(y) = \int_0^1 f(x, y) \, dx = 3(\tfrac{1}{6}+y-y^2)$$

Thus

$$E(X) = \int_0^1 xg(x) \, dx = 2 \int_0^1 (\tfrac{1}{6}x+x^2-x^3) \, dx = \tfrac{1}{2} = E(Y)$$

Also

$$E(XY) = \int_0^1 \int_0^1 xy \, f(x, y) \, dxdy$$
$$= 3 \int_0^1 \int_0^1 [(x^2 y+xy^2)-(x^3 y+xy^3)] \, dxdy = \tfrac{1}{4}$$

Hence $C(X, Y) = E(XY)-E(X)E(Y) = 0$, i.e. X and Y are uncorrelated. However, since $f(x, y) \neq u(x)v(y)$, the variables are not independent. □

Problem 3.16 AOB is the diameter of a semi-circle with centre O and radius a. A point P is chosen at random on the circumference such that the angle POB has the p.d.f. $f(\theta) = 1/\pi$, $0 < \theta < \pi$. Find the coefficient of correlation between AP and BP.

Since $AP^2 + BP^2 = 4a^2$, AP and BP are perfectly related and it can be argued that the coefficient of correlation should be ± 1. Discuss this argument.

Solution. From Fig. 3.1 we have $AP = 2a\cos\frac{1}{2}\theta$, $BP = 2a\sin\frac{1}{2}\theta$, where θ is a r.v. with p.d.f. $f(\theta) = 1/\pi$, $0 \leqslant \theta \leqslant \pi$. Thus

$$E(AP) = \int_0^\pi \frac{2a}{\pi}\cos\frac{\theta}{2}\,d\theta = \frac{4a}{\pi} = E(BP) \qquad \text{from symmetry}$$

$$E(AP^2) = \int_0^\pi \frac{4a^2}{\pi}\cos^2\frac{\theta}{2}\,d\theta = 2a^2 = E(BP^2)$$

$$V(AP) = V(BP) = 2a^2(\pi^2 - 8)/\pi^2.$$

$$E(APBP) = \int_0^\pi \frac{4a^2}{\pi}\cos\frac{\theta}{2}\sin\frac{\theta}{2}\,d\theta = \frac{4a^2}{\pi}$$

$$C(AP, BP) = 4a^2(\pi - 4)/2,$$

so that the correlation coefficient ρ is $2(\pi - 4)/(\pi^2 - 8)$.

Although this value is close to -1, the correlation coefficient, measuring only the degree of linear relationship, does not fully measure the relationship actually present. □

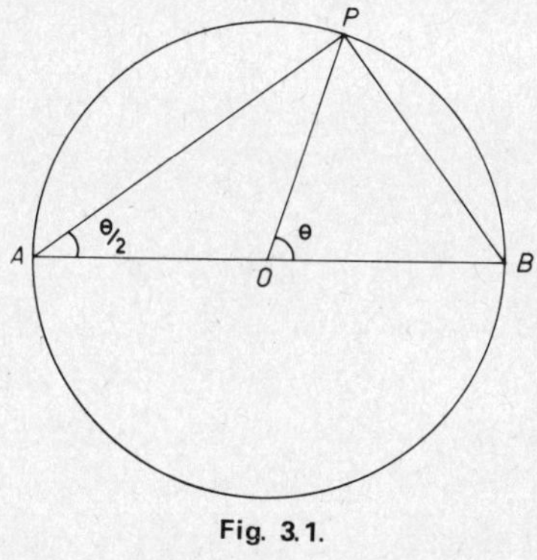

Fig. 3.1.

3.4.3 Generating functions As in the univariate case, discussed in Section 2.4.3, bivariate probability generating function, moment generating function and characteristic function are very useful; the identification of the probability distribution or the probability density function of a random variable is often achieved through one of these functions. They are used for computing various characteristics, for example, the moments of a random variable.

The probability generating function $G(z_1, z_2)$ is particularly useful for a discrete random variable (R, S), where R and S take integral values only. We have in general

$$G(z_1, z_2) = E(z_1^X z_2^Y).$$

For the particular case of an integral-valued random variable (R, S) where $R \geqslant 0$, $S \geqslant 0$, we have

$$G(z_1, z_2) = \sum_{s=0}^{\infty} \sum_{r=0}^{\infty} z_1^r z_2^s P(r, s).$$

Some of the properties of $G(z_1, z_2)$ are:

1. $G(1, 1) = 1$
2. $G(z_1, 1) = G(z_1) = E(z_1^R)$
3. $G(1, z_2) = G(z_2) = E(z_2^S)$
4. $E(R) = \partial G/\partial z_1|_{z_1=1,\, z_2=1}$
5. $E(S) = \partial G/\partial z_2|_{z_1=1,\, z_2=1}$
6. $E(RS) = \partial^2 G/\partial z_1\, \partial z_2|_{z_1=1,\, z_2=1}$
7. $E[R(R-1)] = \partial^2 G/\partial z_1^2|_{z_1=1,\, z_2=1}$. Since $V(R) = E[R(R-1)] + E(R) - [E(R)]^2$, we see that, as in the univariate case, $\partial^2 G/\partial z_1^2$ leads to $V(R)$.
8. $E[S(S-1)] = \partial^2 G/\partial z_2^2|_{z_1=1,\, z_2=1}$.

Note that given a $G(z_1, z_2)$ which can be expanded, we can find $P(r, s)$ as the coefficient of $z_1^r z_2^s$.

The moment generating function, $M(t_1, t_2)$, is given by

$$M(t_1, t_2) = E[e^{(t_1 X + t_2 Y)}].$$

The properties of $M(t_1, t_2)$ are similar to those of $G(z_1, z_2)$ when we evaluate $M(t_1, t_2)$ and its derivatives at $t_1 = 0$, $t_2 = 0$. For example:

1. $M(0, 0) = 1$
2. $M(t_1, 0) = M(t_1)$
3. $E(X) = \partial M/\partial t_1|_{t_1=0,\, t_2=0}$

The characteristic function, as in the case of a univariate random variable is given by $E[e^{i(t_1 X + t_2 Y)}]$.

3.4.4 Some useful properties for a sum of random variables Given k random variables X_j, $j = 1, 2, 3, \ldots, k$, we have

1. $V\left(\sum_{j=1}^{k} a_j X_j\right) = \sum_{j=1}^{k} a_j^2 V(X_j) + \sum_{\substack{m=1 \\ m \neq n}}^{k} \sum_{n=1}^{k} a_m a_n C(X_m, X_n)$. In particu-

lar if $\{X_i\}$ are independent so that $C(X_m, X_n) = 0$ for $m \neq n$, and if $a_j = \pm 1$ for $j = 1, 2, \ldots, k$, we have $V(\sum_j a_j x_j) = \sum_j V(X_j)$.

2. If $\{X_j\}$ are independent and $Y = \sum_{j=1}^{k} X_j$,

(a) $E(z^Y) = E(z^{X_1})E(z^{X_2})\ldots E(z^{X_k})$ or $G_Y(z) = \prod_{j=1}^{k} G_{X_i}(z)$.

(b) $M_Y(t) = \prod M_{X_j}(t)$.

(c) $E(e^{itY}) = E(e^{itX_1})E(e^{itX_2})\ldots E(e^{itX_k})$.

Problem 3.17 Find the probability generating function, $G(z_1, z_2)$, of the random variable (X, Y) of Problem 3.2. Hence verify that $E(Y) = \frac{1}{3}$.

Solution.

$$G(z_1, z_2) = E(z_1^X z_2^Y) = \sum_x \sum_y z_1^x z_2^y P(x, y).$$

Hence

$$G = (z_1^2/36)(1+2z_1+3z_1^2+4z_1^3+5z_1^4+4z_1^5+3z_1^6+2z_1^7+z_1^8) \\ +(2z_2 z_1^7/36)(1+z_1+z_1^2+z_1^3+z_1^4)+z_2^2 z_1^{12}/36.$$

Note that $G(1, 1) = \frac{25}{36}+\frac{10}{36}+\frac{1}{36} = 1$.

$$E(Y) = \partial G(1, 1)/\partial z_2 = \tfrac{10}{36}+\tfrac{2}{36} = \tfrac{1}{3}.$$ □

Problem 3.18 Five unbiased dice are thrown. Find the probability that the sum of the scores is 12.

Solution. Let R_i be a r.v. representing the outcome of the throw of the ith dice. Then $\{R_i\}$ are independent identically distributed r.v.s. with

$$P(R_i = r) = \tfrac{1}{6}, \qquad r = 1, 2, 3, 4, 5, 6 \text{ and p.g.f.}$$
$$\tfrac{1}{6}(z^1+z^2+\ldots+z^6) = (1-z^6)/6^z(1-z).$$

If R is the sum of the scores, we have $R = R_1+R_2+R_3+R_4+R_5$ and hence the p.g.f. of R is

$$G(z) = 6^{-5}z^5(1-z^6)^5(1-z)^{-5}.$$

To find $P(X = 12)$ we identify the coefficient of z^{12} in $G(z)$. Expanding by the binomial theorem, the p.g.f. may be written as

$$\frac{1}{6^5}z^5(1-5z^6+\ldots)(1+5z+\frac{5.6}{1.2^{z^2}}+\ldots+\frac{8.9.10.11}{1.2.3.4}z^7+\ldots)$$

Hence

$$P(X = 12) = \frac{1}{6^5}\left[\frac{8.9.10.11}{1.2.3.4} - 25\right] = \frac{305}{776} \qquad \square$$

Problem 3.19 Each of the n independent random variables R_i, $i = 1, 2, \ldots, n$, takes the value 1 with probability p and the value 0 with probability $1-p$. Find the probability that $R = R_1+R_2+\ldots+R_n = r$, $r = 0, 1, 2, \ldots, n$.

Solution. Let $q = 1-p$. Then $G_{R_i}(z) = z^0q+zp = q+pz$. Since R_i are independent, we have

$$G_R(z) = (q+pz)^n = \sum_0^n \binom{n}{r} p^r q^{n-r} z^r.$$

Hence $P(r) =$ coefficient of $z^r = \binom{n}{r} p^r(1-p)^{n-r}$. $\square$

Problem 3.20 A random variable X has the p.d.f.

$$\begin{aligned} f(x) &= x, & 0 < x < 1 \\ &= 2-x, & 1 < x < 2. \end{aligned}$$

Find the moment generating function of X. Also find the moment generating function of a random variable W which has the p.d.f. $g(w) = 1$, $0 < w < 1$. What can be deduced about the p.d.f. of the sum of two independent variables like W?

Solution.

$$\begin{aligned} M_X(t) &= \int_0^2 e^{xt} f(x)\, dx = \int_0^1 xe^{xt}\, dx + \int_1^2 (2-x)e^{xt}\, dx \\ &= (e^t-1)^2/t^2 \\ M_W(t) &= \int_0^1 e^{wt}\, dw \\ &= (e^t-1)/t. \end{aligned}$$

Therefore $M_X(t) = [M_w(t)]^2$, which means that the sum of two independent random variables like W behaves like the random variable X. $\square$

Problem 3.21 The p.d.f. of a continuous random variable (X, Y) is $f(x, y) = 2\exp[-(x+y)]$, $x > 0$, $y > x$. Find the moment generating function and hence find the correlation coefficient.

Solution.

$$\begin{aligned} M(t_1, t_2) &= E[\exp(Xt_1+Yt_2)] = \int_0^\infty \int_x^\infty 2\exp(xt_1+yt_2)\exp(-x-y)dydx \\ &= 2/(1-t_2)(2-t_1-t_2) \end{aligned}$$

Note that $M(0,0) = 1$. Now

$$E(X) = \partial M(0,0)/\partial t_1 = \tfrac{1}{2}, \qquad E(Y) = \partial M(0,0)/\partial t_2 = \tfrac{3}{2},$$
$$E(XY) = \partial^2 M(0,0)/\partial t_1\,\partial t_2 = 1, \qquad E(X^2) = \partial^2 M(0,0)/\partial t_1^2 = \tfrac{1}{2},$$
$$E(Y^2) = \partial^2 M(0,0)/\partial t_2^2 = \tfrac{7}{2}.$$

Hence $\quad C(X,Y) = E(XY) - E(X)E(Y) = \tfrac{1}{4},$

$$V(X) = E(X^2) - [E(X)]^2 = \tfrac{1}{4}, \qquad V(Y) = \tfrac{5}{4}.$$

so that $\quad \rho = C(X,Y)/\sqrt{[V(X)V(Y)]} = 1/\sqrt{5}.$ □

Problem 3.22 Peter has three candles i, $i = 1, 2, 3$. The time in hours for which the ith candle will burn is a random variable X_i with p.d.f. $f(x) = (2/a_i^2)x$, $0 < x < a_i$. Peter plans to use the candles, one by one, in the event of a power cut. Find the mean and the variance of the total time for which the three candles will burn.

Solution. A reasonable assumption is that X_i are independent. We are interested in $E(X_1 + X_2 + X_3)$ and $V(X_1 + X_2 + X_3)$. Now,

$$E(X_i) = \int_0^{a_i} (2/a_i^2)x^2\,dx = 2a_i/3.$$

Hence

$$E(X_1 + X_2 + X_3) = \tfrac{2}{3}(a_1 + a_2 + a_3).$$

Also

$$E(X_i^2) = \int_0^{a_i} (2/a_i^2)x^3\,dx = \tfrac{1}{2}a_i^2,$$

so that

$$V(X_i) = E(X_i^2) - [E(X)]^2 = a_i^2/18.$$

Therefore $V(X_1 + X_2 + X_3) = \frac{1}{18}(a_1^2 + a_2^2 + a_3^2)$, since X_i are independent. □

Problem 3.23 Random variable X_1, X_2 are independent. If $Y = aX_1 + bX_2$ and $W = aX_1 + cX_2$, where a, b are known constants, find c such that Y, W are uncorrelated.

Solution.

$$E(Y) = a\,E(X_1) + b\,E(X_2), \qquad E(W) = a\,E(X_1) + c\,E(X_2),$$
$$E(YW) = E(a^2X_1^2 + abX_1X_2 + acX_1X_2 + bcX_2^2)$$
$$= a^2E(X_1^2) + (ab + ac)E(X_1)E(X_2) + bc\,E(X_2^2).$$

Hence

$$C(Y,W) = E[YW - E(Y)E(W)] = a^2V(X_1) + bc\,V(X_2)$$
$$= 0 \quad \text{if } c = -a^2V(X_1)/[b\,V(X_2)]$$ □

Problem 3.24 A mixed tennis doubles tournament attracts n husbands and their wives. The pairings are done at random and R represents the number of 'correct' pairs, i.e. a husband and his wife paired together. Find $E(R)$ and $V(R)$.

Solution. Let $\{X_i\,;\, i = 1,\ldots,n\}$ be such that:

$$X_i = 1 \text{ if the } i\text{th pair is 'correct'},$$
$$= 0 \text{ otherwise.}$$

Then $R = X_1+\ldots+X_n$. Also $P(X_i = 1) = 1/n$, and $P(X_i = 0) = 1-1/n$. Hence

$$E(X_i) = \frac{1}{n}, \qquad V(X_i) = \frac{1}{n}\cdot\frac{n-1}{n},$$

so that
$$E(X_i X_j) = \frac{1}{n}\cdot\frac{1}{n-1}, \qquad i \neq j;$$

$$C(X_i, X_j) = \frac{1}{n(n-1)} - \left[\frac{1}{n}\right]^2 = \frac{1}{n^2(n-1)}, \qquad i \neq j.$$

Hence
$$E(R) = \sum_{i=1}^{n} E(X_i) = 1,$$

$$V(R) = \sum_{i=1}^{n} V(X_i) + \sum_{i=1}^{n}\sum_{\substack{j=1\\ i\neq j}}^{n} C(X_i X_j),$$

$$= \frac{n-1}{n} + n(n-1)\left[\frac{1}{n^2(n-1)}\right] = 1. \qquad \square$$

3.4.5 Conditional expectation We have seen in sections 3.2 and 3.3 that given a random variable (X, Y) and the information that Y has taken a particular value y, the conditional probability $P(x/y)$ (or the conditional probability density function $u(x \mid y)$) is of importance. These conditional probability functions may be used to compute conditional expected values. For example

$$E(X \mid y) = \sum_{i=1}^{\infty} x_i P(x_i \mid y), \qquad \text{(discrete case)}$$
$$= \int_{-\infty}^{+\infty} x\, u(x \mid y)\, dx, \qquad \text{(continuous case).}$$

Now $E(X \mid y)$ would, in general, be a function of y and so would assume different values as Y assumes different values y. We can thus regard $E(X \mid y)$ as a random variable and compute, for example, its expected value. A convenient notation for this operation is $E_Y[E(X \mid Y)]$, showing that we are taking the expected value of $E(X \mid y)$ over the distribution of Y. Some useful results are

$$\text{(i) } E_Y[E(X \mid y)] = E(X), \qquad \text{(ii) } E_X[E(Y \mid x)] = E(Y),$$
$$\text{(iii) } E(XY) = E[XE(Y \mid x)] = E[Y\,E(X \mid y)].$$

In general if $h(X)$ is a function of X, the conditional expected value of $h(X)$ is given by

$$E[h(X) \mid y] = \sum_{i=1}^{\infty} h(x_i)P(x_i \mid y) \qquad \text{(discrete case)}$$
$$= \int_{-\infty}^{+\infty} h(x)u(x \mid y)\,dx \qquad \text{(continuous case)}$$

Problem 3.25 A continuous random variable (X, Y) has the p.d.f. $f(x, y) = e^{-x}$, $x > y > 0$. Find $E(X \mid y)$, $E(Y \mid x)$, $V(Y \mid x)$.

Solution. The marginal densities are

$$u(x) = \int_0^x f(x, y)\,dy = \int_0^x e^{-x}\,dy = xe^{-x}.$$
$$v(y) = \int_y^{\infty} f(x, y)\,dx = \int_y^{\infty} e^{-x}\,dx = e^{-y},$$

so that

$$u(x \mid y) = f(x, y)/v(y) = e^{-(x-y)}, \qquad x > y,$$
$$v(y \mid x) = f(x, y)/u(x) = 1/x, \qquad 0 < y < x.$$

Now

$$E(X \mid y) = \int_y^{\infty} x\,u(x \mid y)\,dx = (y+1)$$
$$E(Y \mid x) = \int_0^x y/x\,dy = \tfrac{1}{2}x$$
$$E(Y^2 \mid x) = \int_0^x y^2/x\,dy = \tfrac{1}{3}x^2$$
$$V(Y \mid x) = E(Y^2 \mid x) - [E(Y \mid x)]^2 = x^2/3 - x^2/4 = x^2/12. \ \square$$

Problem 3.26 Peter has n salesmen, and the probability that during a sales campaign exactly r of the n salesmen will be successful depends on the 'state of the market'. This state of the market has been characterized as a random X with the p.d.f. $f(x) = (m+1)(1-x)^m$, $0 < x < 1$. If the probability of r successful salesman, given the state of the market is $P(Y \mid x) = \binom{n}{r} x^r(1-x)^{n-r}$, find the expected number of successful salesmen.

Solution

$$E(R \mid x) = \sum_0^n r\binom{n}{r} x^r(1-x)^{n-r} = nx.$$

Thus $$E(R) = E_X[E(R \mid x)] = E_X(nx)$$
$$= n\int_0^1 (m+1)x(1-x)^m\,dx = n/(m+2). \qquad \square$$

***Problem* 3.27** For a random variable (X, Y), the quantities $E(X)$, $V(X)$, $E(Y)$, $V(Y)$ and the correlation coefficient ρ are known. If $E(Y \mid x) = \alpha+\beta x$, where α, β are unknown, find α, β.

Solution. $E(Y \mid x) = \alpha+\beta X$ so that
$$E_X[E(Y \mid x)] = E(Y) = E_X[\alpha+\beta x] = \alpha+\beta E(X).$$
Hence
$$\alpha = E(Y)-\beta E(X)$$
Now $E_X[X\,E(Y \mid x)] = E(XY) = E_X[\alpha x+\beta x^2]$ and so
$$E(XY) = \alpha E(X)+\beta E(X)^2.$$

We now get
$$C(X, Y) = \beta V(X)$$

so that $$\beta = C(X, Y)/V(X) = \rho\sqrt{[V(Y)/V(X)]}$$
Thus α, β have been evaluated. Note that we can write
$$E(Y \mid x) = E(Y)-\beta[x-E(x)] \qquad \square$$

EXERCISES

1. A random variable X takes the values $x = 1, 2, 3, 4$ and 5, each with probability 0·2. Two independent observations on X are made and the random variable R is the number of even values observed and the random variable S is the total of the values observed. Find (i) $P(S = 6 \mid R = 0)$, (ii) $E(R \mid S = 6)$, (iii) $V(R \mid S = 6)$.

2. A random variable X has the probability density function $f(x) = \lambda e^{-\lambda x}$, $x > 0$. In a sample of n random observations on X, the number of observations in $(0, 1/\lambda)$ and $(1/\lambda, 2/\lambda)$, denoted by R and S, are noted. Find the coefficient of correlation between R, S.

3. Electric light bulbs produced by a manufacturer have life times (in hours) which are independently exponentially distributed with parameter λ. Find the probability that from two bulbs chosen at random, one will last more than twice as long as the other.

4. The conditional probability distribution of a random variable R, given that another random variable X has taken the value x, is

$P(r|x) = \binom{K}{r} x^r(1-x)^{K-r}$, $r = 0, 1, 2, \ldots, K$. The p.d.f. of X is

$$f(x) = \frac{1}{B(m,n)} x^{m-1}(1-x)^{n-1}, \qquad 0 < x < 1,$$

where the constant $B(m,n)$ is given by $B(m,n) = \int_0^1 x^{m-1}(1-x)^{n-1}\,dx$. Find $P(r)$.

5. The number of customers, N, entering a store during a given day is a random variable with mean μ_N, variance σ_N^2 and p.g.f. $G(z)$. The amounts spent by the customers are independent, identically distributed random variables, with mean μ, variance σ^2 and p.g.f. $H(z)$. Obtain (i) the p.g.f., (ii) the mean, and (iii) the variance of the total amount spent on the day.

Chapter 4

Binomial and Poisson Probability Distributions

4.1 Binomial Probability Distribution A binomial probability distribution arises when a series of n independent trials is carried out; each trial has only two possible outcomes, usually labelled 'success' and 'failure'. In each of the n trials, $P(\text{success}) = p$, $P(\text{failure}) = 1-p = q$. The number of successes R is a random variable and $P(R = r) = P(r)$, is easily seen to be

$$P(r) = \binom{n}{r} p^r q^{n-r}, \qquad r = 0, 1, 2, \ldots, n.$$

Note that $P(r)$ can be generated from an expansion of $(q+p)^n$, hence the name *binomial* probability distribution. It is convenient to call R a binomial variate or a binomial random variable, with parameters n, p. Some values of the binomial distribution function are given in Table 1 of the appendix.

It is instructive to regard R as the sum of n independent random variables R_i, $i = 1, 2, \ldots, n$ with $P(R_i = 0) = q$, $P(R_i = 1) = p$. Note that

$$\begin{aligned} E(R_i) &= 0 \times q + 1 \times p = p, \\ E(R_i^2) &= 0 \times q + 1 \times p = p, \\ V(R_i) &= E(R_i^2) - [E(R_i)]^2 = p - p^2 = p(1-p) = pq. \\ E(z^{R_i}) &= z^0 q + zp = q + pz. \end{aligned}$$

Now $R = \sum_{i=1}^{n} R_i$ so that

$$\begin{aligned} E(R) &= E(\sum_i R_i) = \sum_i E(R_i) = np \\ V(R) &= V(\sum_i R_i) = \sum_i V(R_i) \qquad (R_i \text{ independent}) \\ &= npq. \\ G_R(z) &= \prod_i G_{R_i}(z) = (q+pz)^n. \qquad (R_i \text{ independent}) \end{aligned}$$

We can deduce that $P(r) = \binom{n}{r} p^r q^{n-r}$ from the fact that $G(z) = (q+pz)^n$.

With $n = n_1 + n_2$ we have $G(z) = (q+pz)^{n_1+n_2}$. This shows that if R_1, R_2 are independent binomial variates with parameters (n_1, p), (n_2, p) then $R = R_1 + R_2$ is a binomial variate with parameters (n_1+n_2, p).

In many practical problems, a binomial probability distribution is a useful model because $P(r)$ is the limit of an *hypergeometric probability*. An hypergeometric probability distribution arises when we have N items, there being N_1 items of type A and $(N-N_1)$ items of type $\bar{A}$. Let $p = N_1/N$. If a random sample of n items is taken from N items, the probability of getting exactly r items of type A is easily seen to be

$$\binom{N_1}{r}\binom{N-N_1}{n-r}\bigg/\binom{N}{n}.$$

If, now, $N \to \infty$ while N_1/N remains constant at p, the above hypergeometric probability tends to $\binom{n}{r}p^r q^{n-r}$. Thus in practice a binomial probability distribution would be useful for problems involving an hypergeometric situation with large N.

Problem 4.1 An ordinary die with six faces is thrown four times. Find the probability that, (i) a six is not observed, (ii) a six is observed once, and (iii) a six is observed at least twice.

Solution. We have a binomial situation with $n = 4$ and $p = \frac{1}{6}$.

(i) $P(0) = (\frac{5}{6})^4 = 0{\cdot}482$.

(ii) $P(1) = 4\times\frac{1}{6}\times(\frac{5}{6})^3 = \frac{2}{3}(\frac{5}{6})^3 = 0{\cdot}386$.

(iii) $1-(\frac{5}{6})^4-\frac{2}{3}(\frac{5}{6})^3 = 1-\frac{5}{4}(\frac{5}{6})^2 = 0{\cdot}132$ □

Problem 4.2 In a multiple choice test there are ten questions and each question has four answers only one of which is the right one. A candidate's performance is judged to be satisfactory if he gets at least six right answers. An unprepared candidate chooses the answers at random; find the probability that his performance will be judged to be satisfactory.

Solution. We have a binomial situation with $p = 0{\cdot}25$ and $n = 10$. The required probability, P, is

$$P = \sum_{6}^{10}\binom{10}{r}(0{\cdot}25)^r(0{\cdot}75)^{10-r} = 0{\cdot}020.$$ □

Problem 4.3 Ten items are selected at random from a large batch, and the batch is rejected if two or more of the ten items are defective. Find the probability of rejecting the batch if 5% of the items of the batch are defective.

Solution. We treat the probability of a defective as being constant for each of the ten items, the justification being the size of the batch. Thus we have a binomial situation with $n = 10$ and $p = 0{\cdot}05$. The required probability P is:

$$P = 1-P(0)-P(1) = 1-(0{\cdot}95)^{10}-0{\cdot}5\times(0{\cdot}95)^9 = 0{\cdot}086.$$

Note the use of a binomial approximation to a hypergeometric distribution. This example is an application of the result quoted on page 51. □

Problem 4.4 Ten per cent of the words spoken by a politician are 'Er'. How many words must the politician speak such that the probability of at least one 'Er' is at least 0·95?

Solution. We assume that if the politician speaks n words, then this corresponds to a binomial situation with $p = 0{\cdot}10$.

Now $0{\cdot}95 \leqslant P(\geqslant 1) = 1-0{\cdot}9^n$ gives $n \geqslant \ln 0{\cdot}05/\ln 0{\cdot}90$. Thus the required answer is $n = 29$ at least. □

Problem 4.5 Find the value of p such that the variance of a binomial variate is maximized.

Solution. $V(R) = np(1-p)$. Thus $dV/dp = n-2np = 0$ gives $p = \frac{1}{2}$.

Note that $d^2V/dp^2 = -2n < 0$, so that $p = \frac{1}{2}$ gives a maximum value of $V(R)$. □

Problem 4.6 The median of a continuous random variable X is m. In an experiment, n independent observations on X are made. If the number of observations greater than m is c or less, action A will be taken; otherwise action B will be taken. Find the probability that action A will be taken.

Solution. $P(X > m) = P(X \leqslant m) = \frac{1}{2}$. Thus with n independent observations on X, we generate a binomial distribution with $p = \frac{1}{2}$, and so

$$P(\text{action } A) = \sum_{r=0}^{c} \binom{n}{r} (\tfrac{1}{2})^n. \qquad \square$$

Problem 4.7 Peter and Paul play a game in which an ordinary six-sided die is to be thrown n times. Peter will receive £1 from Paul if there is an odd number of sixes and Paul will receive £x from Peter if there is an even number of sixes. Find x for the game to be fair.

Solution. $P(\text{six at a throw}) = \frac{1}{6} = p$, $q = 1-p$. Now

$$P(\text{even no. of sixes}) = \binom{n}{0}q^n + \binom{n}{2}p^2q^{n-2} + \dots$$

$$= \tfrac{1}{2}\{(q+p)^n + (q-p)^n\} = \tfrac{1}{2}\{1+(\tfrac{2}{3})^n\}.$$

Hence $P(\text{odd no. of sixes}) = \frac{1}{2}\{1-(\frac{2}{3})^n\}$.

For a fair game, Peter and Paul should expect to win the same amount, so that

$$\tfrac{1}{2}\{1-(\tfrac{2}{3})^n\} = x \,.\, \tfrac{1}{2}\{1+(\tfrac{2}{3})^n\}$$

$$x = [1-(\tfrac{2}{3})^n]/[1+(\tfrac{2}{3})^n]. \qquad \square$$

Problem 4.8 An electronic device has n transistors and the device will flash a red light if one or more transistors fail. When the red light flashes the device is examined in order to detect the faulty transistors. If during a specified period of operation, each of the n transistors has, independently a probability p of failing, find the probability that exactly r faulty transistors will be detected.

Solution. For detection of faulty transistors, the red light must have flashed; that is, we know that one or more transistors have failed. Thus we have a binomial variate R for which $R > 0$. Now $P(R > 0) = 1 - q^n$, where $q = 1 - p$. Thus the required probability is

$$P(R = r | R > 0) = \frac{\binom{n}{r} p^r q^{n-r}}{1 - q^n}, r = 1, 2, 3 \ldots$$

Comment. This is an example of a truncated binomial variable. □

Problem 4.9 The mean and variance of a binomial variate are μ and $\alpha\mu$ respectively. Find the probability of no successes; one success; at least two successes.

Solution. $\mu = np, \alpha\mu = npq.$

Thus $q = \alpha, p = 1 - \alpha, n = \mu/(1 - \alpha)$

so that $P(0) = q^n = (\alpha)^{\mu/(1-\alpha)}$,

$$P(1) = npq^{n-1} = \mu(\alpha)^{(\mu+\alpha-1)/(1-\alpha)}$$

$$P(\geqslant 2) = 1 - P(0) - P(1).$$ □

Problem 4.10 Independent binomial variates R_i, $i = 1, 2, 3 \ldots k$, have parameters n and p_i. Find the mean and variance of T where $T = \sum_{i=1}^{k} R_i$ and compare these with the mean and variance of a binomial variate R with parameters nk and $p = \sum_{i=1}^{k} p_i/k$.

Solution. Now $T = \sum_i R_i$, where R_i are independent. Hence

$$E(T) = \sum_i E(R_i) = n \Sigma p_i = nkp,$$

$$V(T) = \sum_i V(R_i) = n \Sigma p_i(1 - p_i) = nkp - n \Sigma p_i^2.$$

Also $\quad E(R) = nkp = E(T), \quad \text{and} \quad V(R) = nkp(1 - p).$

Thus $\quad V(T)/V(R) = (1 - \Sigma p_i^2/kp)/(1 - p)$

Now $\sum_i (p_i - p)^2 = \sum_i p_i^2 - kp^2 \geqslant 0,$

so that $\sum p_i^2/kp \geqslant p.$

Hence $[V(T)/V(R)] \leqslant 1.$

Note. It is rather surprising that the variance of T which involves variable values of the probability of success is smaller than the variance of R.

Problem 4.11 Independent binomial variates R_i, $i = 1, 2 \ldots$ have parameters n and p^i. Find the mean and the variance of S where $S = \sum_1^k R_i/k$, and compare these with the mean and the variance of R, which is a binomial variate with parameters n and $\pi = (1/k) \sum p^i$.

Solution.

$$E(S) = \frac{1}{k} \sum E(R_i) = \frac{n}{k} \sum p^i = \frac{np(1-p^k)}{k(1-p)}$$

$$V(S) = \frac{1}{k^2} \sum V(R_i) = \frac{n}{k^2} \sum p^i(1-p^i) = \frac{n\pi(1-p^{k+1})}{k(1+p)}.$$

Hence $E(R) = n\pi = E(S)$ and $V(R) = n\pi(1-\pi)$. Now

$$V(R)/V(S) = (1-\pi)k(1+p)/(1-p^{k+1}) = \lambda$$

Note that $\lambda = 1$ for $k = 1$ (as it should be). For $k \geqslant 2$ and $0 < p < 1$, we have

$$\lambda = (1-\pi)k(1+p)/(1-p)(1+\pi k) > 1$$

Note. In this case S is the mean of R_i and its variance is, as expected, smaller than the variance of R.

Problem 4.12 The probability that a patient suffering from a certain disease will react favourably to a certain treatment is p. The treatment is given to n_1 patients in hospital A and to n_2 patients in hospital B. If the total number of patients who react favourably to the treatment is k, find the probability that r patients from the n_1 patients in hospital A react favourably.

Solution. Let R_1, R_2 be the number of patients who react favourably at hospital A, B respectively. Now R_1, R_2 are binomial variates

$$P(R_1 = r) = \binom{n_1}{r} p^r q^{n-r}, \qquad P(R_2 = r) = \binom{n_2}{r} p^r q^{n-r},$$

$$P(R_1+R_2=k)=\binom{n_1+n_2}{k}p^k q^{n_1+n_2-k}.$$

$$P(R_1=r\,|\,R_1+R_2=k)=\frac{P(R_1+R_2=k\,|\,R_1=r)P(R_1=r)}{P(R_1+R_2=k)}$$

$$=\frac{\binom{n_2}{k-r}p^{k-r}q^{n_2-k+r}\binom{n_1}{r}p^r q^{n_1-r}}{\binom{n_1+n_2}{k}p^k q^{n_1+n_2-k}}$$

$$=\frac{\binom{n_2}{k-r}\binom{n_1}{r}}{\binom{n_1+n_2}{k}}.$$

Note that the conditional distribution of R_1, given R_1+R_2, is hypergeometric. □

4.2 Poisson Probability Distribution In many practical problems, we have a binomial situation with large n (say $n > 20$) and small p (say $p < 0{\cdot}1$). In such cases the following approximation is often useful:

$$\binom{n}{r}pr(1-p)^{n-r} \simeq \frac{(np)^r\, e^{-np}}{r!}$$

A mathematical argument for this approximation is that as $n \to \infty$ and $p \to 0$ in such a way that np is constant at λ:

$$\lim \binom{n}{r}p^r(1-p)^{n-r} = \frac{\lambda^r e^{-\lambda}}{r!}, \qquad r = 0, 1, 2, \ldots.$$

If a random R takes the value r with $P(r) = \lambda^r e^{-\lambda}/r!$, we say that R is a Poisson variate with parameter λ. Note that

1. $E(R) = \sum_{r=0}^{\infty} r\lambda^r e^{-\lambda}/r! = \lambda.$
2. $E[(R)(R-1)] = \sum_{r=0}^{\infty} r(r-1)\lambda^r e^{-\lambda}/r! = \lambda^2.$
3. $V(R) = E(R^2)-[E(R)]^2 = E[(R)(R-1)]+E(R)-[E(R)]^2 = \lambda.$
4. $G(z) = E(z^R) = \sum_{r=0}^{\infty} z^r\lambda^r e^{-\lambda}/r! = e^{\lambda(z-1)}.$

If R_i, $i = 1, 2, \ldots k$ are independent Poisson variates with parameters λ_i, and if $R = \sum_i R_i$, then

$$G_R(z) = \prod_{i-1}^{k} G_{R_i}(z) = \exp\{(z-1)\sum_i \lambda_i\}.$$

Thus R is a Poisson variate with parameter $\sum_{i=1} \lambda_i$. In other words,

$$P(r) = (\sum \lambda_i)^r e^{-\Sigma\lambda_i}/r!$$

A Poisson distribution also arises when events are occurring at random in a continuous medium and we ask for the probability of r events in a unit of the medium. A formal definition of this process, with time as an example of a continuous medium, is as follows:

$$\begin{aligned} P(\text{one event in } t, t+\delta t) &= \alpha\delta t + o(\delta t). \\ P(\text{no event in } t, t+\delta t) &= 1 - \alpha\delta t + o(\delta t). \\ P(\text{two or more events in } t, t+\delta t) &= o(\delta t). \end{aligned}$$

where α is the rate of occurrence of the events, i.e. the mean number of events per unit time. If now we ask for the probability of r events in k units of time, we get

$$P(r) = \lambda^r e^{-\lambda}/r!, \qquad r = 0, 1, 2, \ldots, \qquad \lambda = k\alpha.$$

Some values of the Poisson distribution function are given in Table 2 of the appendix.

Problem 4.13 Two per cent of the items produced by a machine are defective, and the defectives occur at random. Find the probabilities that in a sample of 100 items produced by the machine, there will be 0; 1; 2 or more defective items.

Solution. We have a binomial situation with $n = 100$ and $p = 0{\cdot}02$. We can thus use the Poisson approximation with parameter $m = np$. The required probabilities are:

$$P(0) = e^{-2}, \qquad P(1) = 2e^{-2}, \qquad P(\geqslant 2) = 1 - P(0) - P(1).$$

This use of a Poisson approximation is an application of the result quoted on page 55. □

Problem 4.14 The manager of a supermarket looks at the stock in the store soon after closing time every day and telephones his requirements for the next day, and the ordered goods are delivered early in the morning. Large orders will create congestion in the store and insufficient orders may mean lost sales. For a particular type of tinned soup, the manager has decided that he will accept 0·95 as the minimum probability of meeting the daily demand for the soup. Experience has shown the manager that the daily demand can be taken to be a Poisson variate with mean 11 tins. If at the end of a particular day, the stock of tins left is 2, how many tins should the manager order?

Solution. The starting daily stock should be such that the probability of meeting the demand of the day should be at least 0·95. Thus we have to solve

$$\sum_{0}^{k} 11^r e^{-11}/r! \geqslant 0{\cdot}95$$

From tables of Poisson distribution we find that $k = 17$; the manager should order 15 tins. □

Problem 4.15 Peter receives an average of 5 telephone calls between 9 and 10 a.m. on weekdays, the calls being at random. (i) Find the probability that Peter will receive one or more calls between 9 and 10 a.m. on a certain day, and also the probability of exactly 2 calls between 9 and 9.12 a.m. (ii) Find the probability that during a five-day week there will be exactly two days for which Peter will receive no calls between 9 and 9.12 a.m.

Solution. (i) We have a Poisson situation with parameter 5 calls per 60 minutes. Thus for a 12-minute period we have a Poisson situation with parameter 1, and the required probabilities are:

$$P(\geqslant 1 \text{ call in 60 minutes}) = 1-e^{-5},$$
$$P(2 \text{ calls in 12 minutes}) = \tfrac{1}{2}e^{-1}.$$

(ii) Let $P(0 \text{ calls in 12 minutes}) = e^{-1} = p$. Now during each of the five days there is a constant probability p of 0 calls between 9 and 9.12 a.m. Thus we have a binomial situation with $n = 5$ and $p = e^{-1}$, and the required probability is $\binom{5}{2} p^2(1-p)^3$.

Problem 4.16 A source of liquid is known to contain bacteria, with the mean number of bacteria per cubic millimetre (mm^3) equal to 3. The number of bacteria per unit volume can be taken to be a Poisson variate. Ten 0·5 mm^3 millimetre tubes are filled with the liquid. Find the probability that (a) all the tubes will show growth (i.e. contain at least one bacterium); and (b) exactly 7 tubes will show growth.

Solution. Let R be the number of bacteria per 0·5 mm^3. Now $E(R) = 1{\cdot}5$

and $P(R > 0) = 1-e^{-1\cdot5}$. With 10 tubes we have binomial situation with $n = 10$ and $p = 1-e^{-1\cdot5}$.

Thus the required probabilities are:

$$\text{(a)} \quad p^{10} = (1-e^{-1\cdot5})^{10}$$

$$\text{(b)} \quad \binom{10}{7} p^7(1-p)^3 = 120(1-e^{-1\cdot5})e^{-4\cdot5}$$ □

Problem 4.17 The number of defects in a particular type of television set is a Poisson variate, and the mean number of defects per television set is λ. How many sets must be inspected such that the probability of finding at least one defect is at least $1-\alpha$? Give the probability that the nth set inspected will be the first one to be free from defects.

Solution. P (at least one defect in k sets) $= 1-e^{-k\lambda}$.

Now $1 - e^{-k\lambda} = 1-\alpha$ gives $k = \dfrac{\ln \alpha}{\lambda}$

Thus we would round up $-\ln\alpha/\lambda$ to get an integral value of the number of sets that must be inspected.

For the nth set to be the first one free from defects, we must have one or more defects in each of the previous $(n-1)$ sets. Thus the required probability is $(1-e^{-\lambda})^{n-1}e^{-\lambda}$. □

Problem 4.18 A company receives a type of component in batches, there being 2000 components per batch. Each of the components can be either defective or good, and the company can tolerate some but not too many defectives per batch. Not wishing to spend too much inspection effort, the company would like to determine a sampling inspection scheme in which a sample of n components will be taken at random from a batch. If the sample contains c or less defectives, the batch will be accepted without any further inspection; if the sample contains $c+1$ or more defectives, this will be taken as evidence for a batch with too many defectives and the remaining $2000-n$ components will be inspected. Find the values of n and c such that the probability of accepting a batch (no further inspection) with 20 defective components is about 0·95 and the probability of accepting a batch with 60 defective components is about 0·10.

Solution. Let $P_a(d)$ be the probability of accepting a batch with d defective components. If $P(r/n, d)$ is the probability of getting r defectives in a sample of n from a batch with d defectives, then

$$P_a(d) = \sum_{r=0}^{c} P(r/n, d) = \sum_{0}^{c} \binom{d}{r}\binom{2000-d}{n-r}\Big/\binom{2000}{n}.$$

We now have to solve for n and c such that

$$P_a(20) \simeq 0{\cdot}95 \qquad \text{and} \qquad P_a(60) \simeq 0{\cdot}10.$$

Clearly this is computationally a tedious problem; it requires a 'trial and error' solution. The computational difficulties can be considerably reduced by exploiting the fact that (see page 51.)

$$P(r/n, d) \simeq \binom{n}{r} p^r(1-p)^{n-r}, \qquad \text{where} \qquad p = \frac{d}{2000}$$

A further simplification is possible in this case where $p = 0{\cdot}01$ for $d = 20$, and $p = 0{\cdot}03$ for $d = 60$; the values of p are small and the Poisson approximation can be tried even though the value of n is not, at the moment, known. With the Poisson approximation we have to solve the equations

$$\sum_{r=0}^{c} (0{\cdot}01n)^r \, e^{-0{\cdot}01}/r! \simeq 0{\cdot}95; \qquad \sum_{r=0}^{c} (0{\cdot}03n)^n/r! \simeq 0{\cdot}10.$$

The fact that one of the unknown constants n is a factor of the Poisson parameters simplifies the trial and error procedure. We see that if n, c satisfy the above equations then the two Poisson parameters $\lambda_1 = 0{\cdot}01n$, $\lambda_2 = 0{\cdot}03n$ are such that $\lambda_2/\lambda_1 = 3$. Table 3 of the appendix gives the values of 100λ for given values of c, $F(c)$. For this particular problem we examine the entries of the columns under $F(c) = 0{\cdot}95$ and $F(c) = 0{\cdot}10$ to find the ratio nearest to 3. This leads to the entries 1177 and 398 whose ratio of 2·96 is nearest to 3. This gives $c = 7$. Now $0{\cdot}01n = 3{\cdot}98$ gives $n = 398$, and $0{\cdot}03n = 11{\cdot}77$ gives $n = 392$, so that a reasonable solution would be $n = 400$, $c = 7$.

Comment: This need for judging the quality of a collection of items on the basis of a sample arises in many situations. In a *single sampling scheme* this judgment is made on the basis of one sample from a batch; a *double sampling scheme* has provision of up to two samples. Other possibilities are *multiple sampling scheme* and *sequential sampling scheme* where the items are examined one by one until judgment is made. The importance of this need and the frequency with which it arises has resulted in a considerable amount of work by many people, and this work is usually given the label of *Acceptance Sampling.* The use of a Poisson distribution is quite common in this work. □

Problem 4.19 A sample of 100 items is taken at random from a large batch of items. If the sample contains 2 or more defectives the batch is rejected, otherwise the batch is accepted. Compute the probability of accepting a batch containing 100 p% defective items for a range of values of p, where $0 < p < 1$.

Solution. For a large batch, we suppose that each of the 200 items has the probability p of being defective. We can now use the Poisson approximation with $\lambda = 100p$. Since Table 3 gives values of 100λ, division of the entries of this table corresponding to $c = 1$ by 100 gives the values of $100p$ for the appropriate probability, $F(c)$, of accepting a batch. This leads to:

Probability of accepting a batch	0·95	0·90	0·75	0·50	0·25	0·10	0·05
Percent of defective items in the batch	0·35	0·53	0·96	1·68	2·69	3·89	4·74

Problem 4.20 For the items of Problem 4.19, a batch with 2% or less defective items is considered to be a good quality batch; a batch with 6% or more defective items is considered to be a poor quality batch. Is the sampling procedure a good one? If not, suggest an appropriate procedure.

Solution. It seems reasonable to have a sampling scheme which will accept good quality batches most of the time and reject poor quality batches most of the time. In other words, we would like a high probability of accepting a batch with 2% defective items and a low probability of accepting a batch with 6% defective items. The probabilities computed for Problem 4.19 show that taking a sample of 100 and rejecting the batch if the sample contains 2 or more defective items means a low probability (less than 0·5) of accepting a batch with 2% defective items. Thus this procedure is unsatisfactory. Clearly for an appropriate procedure we need to quantify 'high' and 'low' probabilities. Suppose we decide that we would like

P (accepting a batch/batch has 2% defective items) = 0·95
P (accepting a batch/batch has 6% defective items) = 0·10.

Now we have the situation of Problem 4.18. Using the Poisson approximation, if n is the unknown sample size, we see that the two Poisson parameters are in the ratio $0{\cdot}06n/0{\cdot}02n = 3$. Thus the nearest entries of Table 3 to this ratio of 3 are again 1177 and 398 giving $c = 7$. Since $1177/6 = 196$ and $398/2 = 199$, a reasonable sampling procedure is to take a sample of 200 from a batch and to reject the batch if the sample contains 8 or more defective items. □

Problem 4.21 In a double sampling scheme, a sample of 50 items is taken from a batch, and the batch is accepted if the sample has no defectives. If the sample has three or more defectives, the batch is rejected. If the sample has one or two defectives, a second sample of 100 items is taken and the total number of defectives in the two samples is noted. If this total number of defectives is five or more, the batch is rejected otherwise the batch is accepted. Find the probability of accepting a batch with 1% defective items. What is the average amount of inspection per batch?

Solution. The batch can be accepted in the following mutually exclusive ways.

(1) No defectives in the first sample. In this case the probability of accepting a batch is approximately $e^{-0 \cdot 5} = 0 \cdot 607$ (Table 2).

(2) One defective in the first sample and three or less defectives in the second sample. The probability for this case using Table 2 with $\lambda = 0 \cdot 5$ and $\lambda = 1 \cdot 0$ is approximately $0 \cdot 303 \times 0 \cdot 981 = 0 \cdot 297$.

(3) Two defectives in the first sample and two or less defectives in the second sample. The probability for this case is $0 \cdot 076 \times 0 \cdot 920 = 0 \cdot 1070$.

The probability of accepting a batch with 1% defective items is the total of the above three probabilities which is equal to 0·974.

To compute the average amount of inspection, note that if the batch is accepted or rejected on the basis of the first sample, the amount of inspection is 50 items; otherwise a second sample is taken, making a total of 150 items. Now P (accepting or rejecting on first results) is

$$e^{-0 \cdot 5} + \sum_{r=3}^{\infty} (0 \cdot 5)^r e^{-0 \cdot 5}/r! = 0 \cdot 607 + 0 \cdot 014 = 0621.$$

So that P (taking a second sample) $= 1 - 0 \cdot 621 = 0 \cdot 379$. Thus the average amount of inspection per batch is $50 \times 0 \cdot 621 + 150 \times 0 \cdot 379 = 57 \cdot 9$. Note that in general if n_1, n_2 are the two sample sizes and p_2 is the probability of taking the second sample, the average amount of inspection per batch is $n_1(1-p_2)+(n_1+n_2)p_2 = n_1 + n_2 p_2$. □

Problem 4.22 Peter is in charge of a large army department for cutting the hair of newly acquired recruits. The time, T, in minutes, required for achieving a regulation hair-cut varies from recruit to recruit, and Peter feels that a reasonable model for explaining this variation is to take T to be a random variable with the probability density function $f(t) = \lambda e^{-\lambda t}$. Peter has set a value of $\lambda = \lambda_0$ as the desired (and achieveable performance given willingness) performance for his department. Evidence that λ has taken the value $\lambda_1 (\lambda_1 < \lambda_0)$ or smaller, is a signal that prevailing slackness should be dealt with through a long lecture to the department. In order to make judgments about λ, Peter would like to send in a group of n soldiers to be done from time to time and note the number who are not done within three minutes. Peter feels that each group of n sent in should have a probability 0·05 of leading to a lecture if $\lambda = \lambda_0$ and that this probability should be 0·95 if $\lambda = \lambda_1$. Suggest a decision rule for Peter.

Solution. Let p be the probability that a soldier will not be done within three minutes. Now given n soldiers, the probability that exactly r will not be done is

$$P(r) = \binom{n}{r} p^r (1-p)^{n-r}.$$

A reasonable decision rule for Peter is to resort to a lecture if the number of soldiers not done is sufficiently large. Thus Peter would like

$$\sum_{r=c+1}^{n} \mathrm{P(r)} = 0{\cdot}05 \qquad \text{if} \qquad \lambda = \lambda_0,$$

and

$$\sum_{r=c+1}^{n} P(r) = 0{\cdot}95 \qquad \text{if} \qquad \lambda = \lambda_1.$$

Or, equivalently, Peter would like:

$$\sum_{r=0}^{c} \mathrm{P(r)} = 0{\cdot}95 \qquad \text{if} \qquad \lambda = \lambda_0,$$

and

$$\sum_{r=0}^{c} \mathrm{P(r)} = 0{\cdot}05 \qquad \text{if} \qquad \lambda = \lambda_1$$

Now $p = \int_3^\infty \lambda e^{-\lambda t}\, dt = e^{-3\lambda}$, so that

$$P(r) = \binom{n}{r}(e^{-3\lambda_0})^r (1-e^{-3\lambda_0})^{n-r} \qquad \text{if} \qquad \lambda = \lambda_0,$$

and

$$P(r) = \binom{n}{r}(e^{-3\lambda_1})^r (1-e^{-3\lambda_1})^{n-r} \qquad \text{if} \qquad \lambda = \lambda_1.$$

We now have the situation of Problem 4.18 and if the Poisson approximation can be used, the ratio $e^{-3\lambda_1}/e^{-3\lambda_0}$ would be used in searching through Table 3. □

Problem 4.23 Two independent random variables R_1 and R_2 are Poisson variates with parameters λ_1 and λ_2 respectively. Find the probability that $R_1 = r$ given that $R_1 + R_2 = k$, where $0 \leqslant r \leqslant k$.

Solution.

$$P(R_1 = r \mid R_1 + R_2 = k) = \frac{P(R_1 = r)P(R_1 + R_2 = k/R_1 = r)}{P(R_1 + R_2 = k)}$$

Now R_1+R_2 is a Poisson variate with parameter $\lambda_1+\lambda_2$. Also if R_1 is given to be r then $R_2 = k-r$ will result in $R_1+R_2 = k$. Thus the required probability P is

$$P = \binom{k}{r} p^r q^{k-r}, \qquad p = \frac{\lambda_1}{\lambda_1+\lambda_2}, \quad q = \frac{\lambda_2}{\lambda_1+\lambda_2} \qquad \square$$

Note that the conditional distribution of R_1, given R_1+R_2, is binomial.

Problem 4.24 A Poisson variate R has parameter λ. Find the conditional expectation of Z^R given that $R > 0$ and hence find the conditional mean and variance of R given that $R > 0$.

Solution. $P(R = r/R > 0) = \lambda^r e^{-\lambda}/r!(1-e^{-\lambda})$, $r = 1, 2, 3, \ldots$.
Let $H(z) = E(z^R/R > 0)$.

$$H(z) = \frac{1}{1-e^{-\lambda}} \sum_{1}^{\infty} \frac{z^r \lambda^r e^{-\lambda}}{r!} = \frac{e^{-\lambda}(e^{z\lambda}-1)}{1-e^{-\lambda}}$$

Now $E(R \mid R > 0) = H'(1) = \lambda/(1-e^{-\lambda})$ and

$$V(R \mid R > 0) = H''(1)+H'(1)-[H'(1)]^2$$
$$= \frac{\lambda}{1-e^{-\lambda}}\left[1-\frac{\lambda e^{-\lambda}}{1-e^{-\lambda}}\right] \qquad \square$$

Problem 4.25 Random variables X_i, $i = 0, 1, 2, \ldots$ are independent and $E(X_i) = i$. A Poisson variate R is independent of X_i and $E(R) = \lambda$. Find $E\{\sum_0^R X_i\}$.

Solution. Let $T = \sum_0^R X_i$. Now $E(T \mid R = r) = E(\sum_0^r X_i) = r(r+1)/2$, and so

$$E(T) = \sum_{r=0}^{\infty} \frac{r(r+1)}{2} \frac{\lambda^r e^{-\lambda}}{r!} = \frac{\lambda(\lambda+2)}{2}. \qquad \square$$

EXERCISES

1. A random variable X has probability density function $f(x) = [(m+1)/a^{m+1}]\, x^m$, $0 < x < a$. Find the probability that in four independent observations on X exactly two will be less than $\frac{1}{2}a$.

2. Ten items are selected at random from a large batch. The batch is accepted if none of the ten items is defective, and if two or more are defective the batch is rejected. If one of the ten items is defective, another sample of ten is obtained, and the batch is rejected if the total number of

defectives in the twenty items is two or more; otherwise the batch is accepted. Find the probability of accepting the batch if 5% of the items in the batch are defective.

3. A binomial variate R has parameters p, n; the parameter p takes one of two values p_1 and p_2. The probability of p_1 is π and the probability of p_2 is $1-\pi$. Find the probability that $R = n$ given that $R > 0$.

4. An electronic device has two independent switches in series and the device will operate only when both switches function. The probability that a switch will fail is θ. Four such devices have been produced; find the probability that exactly three will operate.

5. A Poisson variate R is such that $P(R = 4) = 0{\cdot}25\, P(R = 3)$. Find $P(R > 0)$.

6. The number of bubbles in a glass sheet is a Poisson variate with parameter. λ. The manufactured glass is inspected and each of the bubbles has, independently, a probability p of being missed by the inspector. Find the probability that a glass sheet leaves the inspector with k undetected bubbles.

Chapter 5

Normal Distribution

5.1 Normal distribution A random variable Y is said to be a normal variate, or is said to be normally distributed, if its probability density function is

$$f(y) = \frac{1}{\sigma\sqrt{(2\pi)}}\exp-\frac{(y-\mu)^2}{2\sigma^2}, \qquad -\infty < y < \infty$$

A normal distribution is very important for several reasons. Experience has shown that it is a reasonable model for explaining the variation in results in a wide variety of situations; in many cases the asymptotic distribution of a random variable is normal; a considerable theory of statistics has been built up on the basis of a normal variate. In view of the frequency with which a normal variate is encountered, it is convenient to have a notation for it and we take Y is $N(\mu, \sigma^2)$ to mean that Y is a normal variate with parameters μ and σ^2. Note that

$$E(Y) = \mu, \qquad V(Y) = \sigma^2$$
$$E(e^{tY}) = \exp(\mu t + \tfrac{1}{2}\sigma^2 t^2)$$
$$E[e^{t(Y-\mu)}] = e^{-t\mu}E(e^{tY}) = \exp(\tfrac{1}{2}\sigma^2 t^2).$$

Thus all the odd moments about μ, that is $E[(Y-\mu)^{2r+1}], r = 0, 1, \ldots$ are 0. (This of course is due to the symmetry of $f(y)$.) Now,

$$E[e^{(aY+b)t}] = \exp[(a\mu+b)t + \tfrac{1}{2}a^2\sigma^2 t^2],$$

so that $(aY+b)$ is $N(a\mu+b, a^2\sigma^2)$, and in particular $X = (Y-\mu)/\sigma$ is $N(0, 1)$. A $N(0, 1)$ variate is usually called a standard or unit normal variate. The simple linear relationship between a $N(\mu, \sigma^2)$ and a $N(0, 1)$ variate makes the extensive and easily available tables of a $N(0, 1)$ variate generally useful. The distribution function for a standard normal variate is given in Table 4 of the appendix.

If $Y_i, i = 1, 2, \ldots, n$ are independent where Y_i is $N(\mu_i, \sigma_i^2)$ and $Z = \sum a_i Y_i$, where a_i are constant, we have

$$E(e^{tZ}) = \exp[t \sum a_i \mu_i + \tfrac{1}{2}t^2 \sum a_i^2 \sigma_i^2]$$

Thus Z is $N(\sum a_i \sigma_i^2)$. For the particular case of $\mu_i = \mu, \sigma_i^2 = \sigma^2$ and $a_i = 1/n$ we get

$$Z = \sum Y_i/n = \overline{Y} \text{ is } N(\mu, \sigma^2/n)$$

Problem 5.1 A random variable Y is $N(\mu, \sigma^2)$. Find $P(\mu - k\sigma < Y < \mu + k\sigma)$ when $k = 1, 2$ and 3.

Solution. $P(\mu - k\sigma < Y < \mu + k\sigma) = P(-k < X < k)$. Thus the required probabilities are 0·6826, 0·9544 and 0·9974.

Note that although Y takes values along the entire real line, most of the probability is contained within, say $\mu \pm 3\sigma$. □

***Problem** 5.2* The length of a component produced by a machine can be taken to be a normal variate. The machine is used for producing components in batches of 10 000 and at the start of a run for producing a batch the machine can be set to yield the required mean length and any one of three values of variance. The required mean is 10 mm, and any length outside the range (9·88, 10·12) means that the component is defective. The machine can be adjusted to produce a variance of 0·0016, 0·0025, 0·0036 at a cost of £800, £500, £50 respectively. If each defective produced means a cost of £1, which variance would you recommend?

Solution. Let the machine be set to yield a mean of 10 mm. Now any length outside (9·88, 10·12) means a defective. Also let c_1 be E (cost due to defectives in a batch), so that $c_1 = 10\,000 \times P$(defective component). We now have:

Variance	0·0016	0·0025	0·0036
P(defective)	0·0028	0·0164	0·0456
c_1	28	164	456
Cost of adjusting	800	500	50
Total cost	828	664	506

Thus a variance of 0·0036 is recommended. □

***Problem** 5.3* Breaking strength of certain welds, in lb per square inch, can be taken to be a normal variate. Process A leads to mean 1100, standard deviation 100; process B leads to mean 1000, standard deviation 40. A weld with a breaking strength less than 900 is a poor weld. Which process should be chosen so as to minimise the per cent of poor welds?

Solution. Let Y be $N(1100, 10000)$, let Z be $N(1000, 1600)$, and let X be $N(0, 1)$.

Now $$P(Y < 900) = P(X < -2) = p_1$$

and $$P(Z < 900) = P(X < -2{\cdot}5) = p_2$$

since $p_2 < p_1$, process B should be chosen. □

***Problem** 5.4* Plugs and sockets are manufactured by independent machines. If P and S are the diameters, in millimeters, of a plug and socket respectively, then for a good fit it is necessary that S is in the range

$(P+0{\cdot}2,\ P+0{\cdot}4)$. Normal variates are good models for explaining the variation in the diameters. If the machines are such that P has mean 1·0 and variance 0·001 and S has mean 1·3 and variance 0·0015, find the percentage of good fits when the plugs and sockets are put together at random.

Solution. For a good fit we require $P+0{\cdot}2 < S < P+0{\cdot}4$; that is, $0{\cdot}2 < S-P < 0{\cdot}4$. Now $E(S-P) = 1{\cdot}3-1{\cdot}0 = 0{\cdot}3$; also $V(S-P) = 0{\cdot}0015+0{\cdot}001 = 0{\cdot}0025$, since S and P are independent. Thus $(S-P)$ is $N(0{\cdot}3,\ 0{\cdot}0025)$ and we find that $P(0{\cdot}2 < S-P < 0{\cdot}4) = 0{\cdot}9545$. We therefore expect about 95·5 per cent good fits. □

Problem 5.5 Breaking strength, as measured in special units, of a type of weld can be taken to be normally distributed. In an experiment involving a very large number of welds, it was seen that 95% of the welds had a breaking strength of more than 900 units and 10% of the welds had a breaking strength of more than 1200 units. Find the mean and the standard deviation of breaking strength.

Solution. Let μ and σ be the mean and the standard deviation of breaking strength Y. Now $(Y-\mu)/\sigma$ is $N(0, 1)$. The number of welds involved in the experiment is very large and we translate the information that 95% of the welds had a breaking strength of more than 900 as:

$$P(Y > 900) = P[X > (900-\mu)/\sigma] = 0{\cdot}95.$$

From Table 5, we see that $P(X > K) = 0{\cdot}95$ if $K = -1{\cdot}645$. Thus we get

$$(900-\mu)/\sigma = -1{\cdot}645.$$

Similarly the information that 10% of the welds exceed a strength of 1200 leads to

$$(1200-\mu)/\sigma = 1{\cdot}282.$$

These two equations are easy to solve and we get $\mu = 1068{\cdot}8$, and $\sigma = 102{\cdot}5$.

□

Problem 5.6 In an assembly operation four tubes, one of type T_1, one of type T_2 and two of type T_3 are joined together 'end-on' as shown in Fig. 5.1.

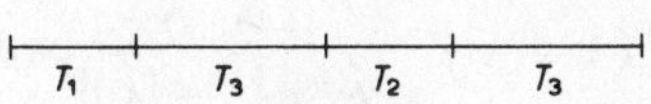

Fig. 5.1.

Tube type T_i is produced by machine i, $i = 1, 2, 3$, and a normal variate is a reasonable approximation for explaining the variability in the length (in mm) of the tubes. The three machines work independently and they lead to the following means and standard deviations.

Machine	*Mean*	*Standard deviation*
1	10	0·5
2	12	0·5
3	20	1·0

The total length of the assembly has to be judged against a specification of 62 ± 3 mm. Find the per cent of assemblies that will fail to meet the specification.

Solution. Let L be the length of the assembled tubes. Now

$$L = T_1 + T_2 + T_3 + T_3 \qquad E(L) = 10+12+20+20 = 62$$

$$V(L) = 0{\cdot}25 + 0{\cdot}25 + 1 + 1 = 2{\cdot}5$$

Thus L is $N(62, 2{\cdot}5)$, so that $X = (L-62)/\sqrt{2{\cdot}5}$ is $N(0,1)$, and

$$P(59 < L < 65) = P(-1{\cdot}90 < X < 1{\cdot}90) = 0{\cdot}9426$$

Hence we expect that about 6% of the lengths will fail to meet the specification. □

Problem 5.7 A random variable is $N(\mu, \sigma^2)$. Find $E(X \mid X \geqslant L)$.

Solution. We consider the p.d.f. of X given that $X \geqslant L$.

$$f(x) = 0, \quad x < L,$$

$$f(x) = \frac{1}{[1-F(L)]\sigma\sqrt{(2\pi)}} \exp - \frac{(x-\mu)^2}{2\sigma^2}, \qquad x \geqslant L.$$

Thus
$$E(X \mid X \geqslant L) = \int_L^\infty \frac{x \exp - [(x-\mu)^2/2\sigma^2]}{[1-F(L)]\sigma\sqrt{(2\pi)}} dx$$

$$= \frac{\mu}{1-F(L)} \int_L^\infty \frac{1}{\sigma\sqrt{(2\pi)}} \exp - \frac{(x-\mu)^2}{2\sigma^2} dx$$

$$+ \frac{\sigma}{\sqrt{(2\pi)}[1-F(L)]} \int_{(L-\mu)^2/2\sigma^2}^\infty e^{-t}\, dt$$

Hence
$$E(X \mid X \geqslant L) = \mu + \frac{\sigma}{\sqrt{(2\pi)}[1-F(L)]} \exp - \frac{(L-\mu)^2}{2\sigma^2}.$$

Note: This is an example of a truncated normal distribution. There are many cases where a truncated distribution is an appropriate model. Note that $E(X \mid X \geqslant L) > \mu$, and one application would be to determine

L such that μ increases by a required amount. This means that we have to solve

$$\frac{\sigma}{\sqrt{(2\pi)}[1-F(L)]}\exp-\frac{(L-\mu)^2}{2\sigma^2} = k. \qquad \square$$

5.2 Normal Approximation for a Binomial Distribution If R is a binomial variate, that is, if $P(r) = \binom{n}{r}p^r q^{n-r}$ where $q = 1-p$, then $(R-np)/\sqrt{(npq)}$ is asymptotically $N(0, 1)$. Thus

$$P(R \leqslant r) = P\left(\frac{R-np}{\sqrt{(npq)}} < \frac{r-np}{\sqrt{(npq)}}\right) \to \int_{-\infty}^{(r-np)/\sqrt{(npq)}} \frac{1}{\sqrt{(2\pi)}} e^{-x^2/2}\, dx \text{ as } n \to \infty$$

In practice the approximation is quite good for moderate values of n, say $n = 15$ if p is in the neighbourhood of 0·5; for p in the neighbourhood of 0 or 1, n has to be larger. The approximation can be improved by 'matching' the discrete values of R with the continuous ones of a normal variate, and a simple correction for continuity is to replace r by $r \pm \frac{1}{2}$ as appropriate. For example, when using the normal approximation we would take:

$$P(R = r) \simeq P(r-\tfrac{1}{2} < Y < r+\tfrac{1}{2})$$
$$P(R \leqslant r) = P(0 \leqslant R \leqslant r) \simeq P(-\tfrac{1}{2} < Y < r+\tfrac{1}{2})$$

where Y is $N(np, npq)$.

***Problem* 5.8** A coin is tossed 20 times and R, the number of heads, is noted. Find $P(R = 15)$, $P(5 \leqslant R \leqslant 15)$.

Solution. We use the normal approximation for the binomial distribution with $n = 20$ and $p = 0{\cdot}5$. Now $E(R) = 10$, $V(R) = 5$. Let Y be $N(10, 5)$. The required probabilities are

$$P(14{\cdot}5 < Y < 15{\cdot}5) = 0{\cdot}0153, \qquad P(4{\cdot}5 < Y < 15{\cdot}5) = 0{\cdot}9862.$$

It is interesting to compare the normal approximations with the binomial probabilities which are:

$$P(R = 15) = 0{\cdot}0148, \qquad P(5 \leqslant R \leqslant 15) = 0{\cdot}9882. \qquad \square$$

***Problem* 5.9** An advertising display uses 200 light bulbs, and the desired effect would be seriously affected if 17 or more bulbs fail during the period of display. If each of the bulbs has independently a probability 0·05 of failing during the period of the display, find the probability that the display will be completed satisfactorily.

Solution. We use the normal approximation for the binomial distribution with $n = 200$ and $p = 0{\cdot}05$. Thus we are concerned with Y, a $N(10, 9{\cdot}5)$ variate. Now

$$P(Y < 16{\cdot}5) = 0{\cdot}9826,$$

which is the required probability.

It is interesting to compare this probability with the answer using a Poisson approximation, which is

$$\sum_{0}^{16} 10^r e^{-10}/r! = 0{\cdot}9730. \qquad \square$$

Problem 5.10 A psychologist has devised an intelligence test for British frogs, and he finds that the score achieved varies from frog to frog. Extensive experimentation suggests that a reasonable model for the variability in the score, S, of British frogs is to take S to be a $N(100, 100)$ variate. The psychologist has now designed a quick test for judging whether or not the score of a frog will exceed 105, as a means for selecting frogs with a score of at least 105. Find the probability that the selection test will result in 150 or more frogs being selected from a group of 500 frogs.

Solution. Now $P(S > 105) = P[(S-100)/10 > 0{\cdot}5] = 0{\cdot}31$. Thus we suppose that each of the 500 frogs have a probability 0·31 of being selected. We therefore have a binomial situation, and the probability of selecting 150 or more frogs is

$$P(R \geqslant 150) = \sum_{r=150}^{500} \binom{500}{r} (0{\cdot}31)^r (0{\cdot}69)^{500-r}.$$

We can use the normal approximation and take R to be approximately $N(135, 93{\cdot}15)$. If now Y is $N(135, 93{\cdot}15)$ then $P(Y > 149{\cdot}5) = 0{\cdot}067$, which is approximately the required probability. $\square$

Problem 5.11 A large international mailing list is held by a computer. The composition of the list by nationality is:

20% of the names are British
12% of the names are French
8% of the names are German
30% of the names are American
10% of the names are African
20% of the names are Asian

The computer has been asked to select 400 names at random from the list. Find the probability that at least 170 European names will be selected.

Solution. We combine the British, French, and German names to get

P (name is European) $= 0{\cdot}40$. We suppose that a European name has a probability $0{\cdot}40$ of being selected at each of the 400 attempts. If R is the number of European names selected, then R is, approximately, a $N(160, 96)$ variate. The required probability is about $0{\cdot}166$. □

5.3 Central Limit Theorem This important theorem explains why a normal variate may be a reasonable model for measurements made in a wide variety of situations. The result is that if Z_i, $i = 1, 2, \ldots$ are independent with $E(Z_i) = \mu_i$, $V(Z_i) = \sigma_i^2$ and if $X = \sum_{i=1}^{n} Z_i$, then under some fairly general conditions the asymptotic distribution of $(X - \sum \mu_i)/\sqrt{\sum \sigma_i^2}$ is $N(0, 1)$. Thus in practice, if n is large we take X to be $N(\sum \mu_i, \sum \sigma_i^2)$. If the value observed when an item is measured is the result of several 'random disturbances' adding up, then we would expect a normal variate to be a reasonable approximation.

Problem 5.12 In a general knowledge test, Peter has to answer 250 questions. For each question Peter has, independently, a probability $0{\cdot}05$, $0{\cdot}8$, $0{\cdot}15$ of scoring 0, 1, 2 respectively. Find the probability that Peter's total score will exceed 290.

Solution

$$E(\text{score per question}) = 0 \times 0{\cdot}05 + 1 \times 0{\cdot}8 + 2 \times 0{\cdot}15 = 1{\cdot}1,$$

$$E(\text{score}^2 \text{ per question}) = 0 \times 0{\cdot}05 + 1 \times 0{\cdot}8 + 4 \times 0{\cdot}15 = 1{\cdot}4,$$

$$V(\text{score per question}) = 1{\cdot}4 - 1{\cdot}21 = 0{\cdot}19,$$

$$E(\text{score for test}) = 250 \times 1{\cdot}1 = 275,$$

$$V(\text{score for test}) = 250 \times 0{\cdot}19 = 47{\cdot}5.$$

Now the probability that a $N(275, 47{\cdot}5)$ variate will exceed 290 is $0{\cdot}015$ and this is, approximately, the required probability. □

Problem 5.13 Peter buys a stock of 200 candles in anticipation of a disruption of the supply of electricity. The time, in minutes, for which a candle will burn is a random variable, with mean 30 and variance 4. The time, in hours, for which Peter will need to burn the candles due to the disruption is a normal variate with mean 80 and variance 574/9. Find the probability that Peter will have some candles to spare when steady supply of electricity is restored.

Solution. Let T, in hours, be the total candle burning time. Thus

$$E(T) = 100, \qquad V(T) = 800/60^2 = \tfrac{2}{9}.$$

Using the central limit theorem we take T to be $N(100, \frac{2}{9})$. Let Y be the time for which candles will be required, so that Y is $N(80, 574/9)$. Some candles will be left if $T > Y$ that is if $T-Y > 0$. Now

$$E(T-Y) = 20, \qquad V(T-Y) = \tfrac{2}{9}+\tfrac{574}{9} = 64.$$

so that $T-Y$ is $N(20, 64)$. Thus the required probability is about 0·99. □

Problem 5.14 A particle moves along the x-axis at times $n, n = 1, 2, 3 \ldots$. If X_n is its position at time n, then $X_n = X_{n-1}+Z_n$, where Z_n are independent and identically distributed random variables with $P(Z_n = 1) = 0{\cdot}5$ and $P(Z_n = -1) = 0{\cdot}5$. Initially the particle is at the origin. Find k such that $P(-k < X_{400} < k) = 0{\cdot}95$.

Solution. Note that $X_n = \sum_{i=1}^{n} Z_i$. $E(Z_i) = 0$, $V(Z_i) = 1$, so that $E(X_n) = 0$, $V(X_n) = n$. Thus, using the central limit theorem, we take X_{400} to be $N(0.400)$. It follows that

$$P(-39{\cdot}2 < X_{400} < 39{\cdot}2) = 0{\cdot}95. \qquad \square$$

5.4 Bivariate normal distribution A random variable (X, Y) is said to be a bivaraite normal variable, or is said to have a bivaraite normal distribution if its probability density function is

$$f(x, y) = \frac{1}{2\pi\sigma_1\sigma_2\sqrt{1-\rho^2}} e^{-Q/2}, \quad -\infty < x < \infty, \quad -\infty < y < \infty,$$

where

$$Q = \frac{1}{(1-\rho^2)} [(x-\mu_1)^2/\sigma_1^2 - 2\rho(x-\mu_1)(y-\mu_2)/\sigma_1\sigma_2 + (y-\mu_2)^2/\sigma_2^2]$$

The marginal probability density functions are given by

$$u(x) = \int_{-\infty}^{+\infty} f(x, y)\, dy = \frac{1}{\sigma_1\sqrt{2\pi}} \exp[-(x-\mu_1)^2/2\sigma_1^2], \quad -\infty < x < \infty$$

$$v(y) = \int_{-\infty}^{+\infty} f(x, y)\, dx = \frac{1}{\sigma_2\sqrt{2\pi}} \exp[-(y-\mu_2)^2/2\sigma_2^2], \quad -\infty < y < \infty.$$

That is, X and Y are $N(\mu_1, \sigma_1^2)$ and $N(\mu_2, \sigma_2^2)$ variates respectively. If X and Y are independent then we have

$$f(x, y) = u(x)\, v(y),$$

and this means that $\rho = 0$. Also if $\rho = 0$ then $f(x, y) = u(x)\, v(y)$. The conditional probability density of Y, given that $X = x$, is

$$v(y/x) = f(x, y)/u(x) = \frac{1}{\sigma_2\sqrt{1-\rho^2}\sqrt{2\pi}} \exp\left[-\frac{1}{2\sigma_2^2(1-\rho^2)} q^2\right],$$

where $q^2 = [y-\{\mu_2+(\rho\sigma_2/\sigma_1)(x-\mu_1)\}]^2$. That is the random variable

Y, when $X = x$, is a $N[\mu_2+(\rho\sigma_2/\sigma_1)(x-\mu_1), \sigma_2^2(1-\rho^2]$ variate. Also the random variable X, when $Y = y$ is a $N[\mu_1+(\rho\sigma_1/\sigma_2)(y-\mu_2), \sigma_1^2(1-\rho^2)]$ variate. Note that $E(X/y)$ and $E(Y/x)$ are linear functions of y and x respectively.

The moment generating function, $M(t_1, t_2)$, is the expected value of $\exp(t_1X+t_2y)$. We have

$$M(t_1, t_2) = \exp[\mu_1 t_1+\mu_2 t_2+(\tfrac{1}{2})(\sigma_1^2 t_1^2+2\rho\sigma_1\sigma_2 t_1 t_2+\sigma_2^2 t_2^2)]$$

Moments of X can be evaluated through computing the partial derivatives with respect to t_1 at the point $t_1 = 0$, $t_2 = 0$. Moments of Y can be similarly computed. Joint moments can be evaluated through differentiation with respect to t_1 and t_2. For example:

$$E(X) = \frac{\partial M(0,0)}{\partial t_1} = \mu_1, \qquad E(X^2) = \frac{\partial^2 M(0,0)}{\partial t_1^2} = \mu_1^2+\sigma_1^2,$$

so that $$V(X) = E(X^2)-[E(X)]^2 = \sigma_1^2 \text{ as before.}$$

$E(XY) = \dfrac{\partial M(0,0)}{\partial t_1\,\partial t_2} = \rho\sigma_1\sigma_2+\mu_1\mu_2$, so that the covariance of (X, Y) is given by

$$C(X, Y) = E(XY)-E(X)\,E(Y) = \rho\sigma_1\sigma_2.$$

Co-efficient of correlation $= C(X, Y)/\sqrt{V(X)\,V(Y)} = \rho, (-1 \leqslant \rho \leqslant 1)$. Thus $\rho = 0$ means that (X, Y) are uncorrelated. Thus in the particular case of a bivariate normal distribution, we see that if X, Y are uncorrelated then X, Y are independent.

Problem 5.15 A random variable (X, Y) has parameters $\mu_1 = 10, \mu_2 = 20$, $\sigma_1^2 = 16$, $\sigma_2^2 = 100$, and ρ. Find k such that
$P[E(Y|x = 15)-k < Y < E(Y|x = 15)+k] = 0{\cdot}95$ when (i) $\rho = -0{\cdot}5$, (ii) $\rho = 0{\cdot}1$, (iii) $\rho = 0{\cdot}9$.

Solution. From Table 5, we deduce that if Z is $N(\mu, \sigma^2)$ then $P(\mu-k < Z < \mu+k) = 0{\cdot}95$ if $k = 1{\cdot}96\sigma$. Thus the required values of k are

$$\text{(i)}\quad k = 1{\cdot}96\times\sqrt{75} = 16{\cdot}97$$
$$\text{(ii)}\quad k = 1{\cdot}96\sqrt{99} = 19{\cdot}51$$
$$\text{(iii)}\quad k = 1{\cdot}96\sqrt{19} = 0{\cdot}86$$

Note that nearer ρ^2 is to 1, smaller is the range about $E(Y/x)$ which contains Y with a probability 0·95. □

Problem 5.16 A psychologist has developed personality indices for Man and his Dog. Experience has suggested that a bivariate normal random variable (X, Y), where X is Man's index and Y is his Dog's index, is a reasonable model for these indices. The parameters for Englishmen and their Dogs are $\mu_1 = 50$, $\mu_2 = 80$, $\sigma_1^2 = 25$, $\sigma_2^2 = 16$, $\rho = 0{\cdot}8$. A Dog has lost his Man, and it is felt that information about the personality index of the Man will help in the search for the Man. If the Dog's index is 70, find the expected value of the Man's index. What is the probability that the Man's index will exceed 50.

Solution. We have $E(X/Y = 70) = 50+(0{\cdot}8)(5/4)(70–80) = 40$. Also $\sigma_1^2(1-\rho^2) = 25\times 0{\cdot}36$, so that $(X/Y = 70)$ is $N(40, 9)$. Thus the probability that the Man's index will exceed 50 is about 0·0005. □

EXERCISES

1. The length $L$of a glass rod produced by an automatic process can be taken to be a normal variate. The specification is that the rods must be between 10 and 11 millmeters. The automatic process rejects 1% of the rods because $L < 10$ and it rejects 2·5% because $L > 11$. Find $E(L)$ and $V(L)$.

2. Independent random variables X_i, $i = 1, 2$, are normal with mean i and variance i^2. Find the probability that $2X_1+4 > \frac{1}{2}X_2+10$.

3. If X is $N(0, 1)$, show that

$$E(X \mid a < X < b) = \frac{f(a)-f(b)}{F(b)-F(a)}.$$

4. Peter has decided to save a small amount every day. If the probabilities of Peter saving 1p, 2p, 5p, each day are 0·2, 0·5, 0·3 respectively, find the probability that Peter will save more than £9·00 in a year (365 days).

Chapter 6

Some Other Distributions

6.1 Geometric and negative binomial distributions A geometric distribution arises when a series of independent trials is carried out; at each trial a defined A has a probability p of occurring and a probability $q = 1-p$ of not occurring and the series of trials is continued till A occurs. The number of necessary trials, R, is a geometric random variable with

$$P(R = r) = q^{r-1}p, r = 1, 2, 3, \ldots$$
$$F(r) = P(R \leqslant r) = 1-q^r.$$

The probability generating function of R is

$$G(z) = \sum_{1}^{\infty} z^r q^{r-1} p = pz/(1-qz),$$
$$E(R) = 1/p,$$
$$V(R) = q/p^2.$$

The geometric random variable has an interesting property; suppose the series of trails has continued up to trial number i, and we now compute the probability that at least j further trials will be necessary till the occurrence of A.

$$P(R \geqslant j+i | R > i) = q^{j+i-1}/q^i = q^{j-1},$$

which is independent of i. A geometric random variable has, thus, a lack of memory; it does not remember the number of trials that have already been carried out when we consider the probability that at least j further trials will be necessary. Problem 2.6 shows another aspect of this lack of memory.

Problem 6.1 Peter and Paul shoot independently until each has hit his own target. Peter, Paul have probabilities 2/5, 2/3 of hitting the targets at each shot respectively. Find the probability that Peter will require less shots than Paul.

Solution. Let X, Y be the number of shots fired by Peter, Paul respectively. Since X, Y are independent we have

$$P(x, y) = (2/5)(3/5)^{x-1}(2/3)(1/3)^{y-1}, x, y = 1, 2, 3 \ldots$$

The required probability, Q, is given by

$$Q = \sum_{x=1}^{\infty} P(X = x, y > x) = \sum_{1}^{\infty} (2/5)(3/5)^{x-1}(1/3)^x = 1/6. \quad \square$$

Problem 6.2 A sample of 50 items is taken every hour from the output of a machine. If the sample contains two or more defective items, the machine is stopped for adjustment. If the machine is producing 2% defective items, find the average interval of time between successive adjustments.

Solution. Each of the 50 sampled items has a probability 0·02 of being defective. The number of defectives in the sample, X, is thus a binomial variate and in this particular case we can use the Poisson approximation. Since $50 \times 0{\cdot}02 = 1{\cdot}0$, we have

$$P(X \geqslant 2) = 1 - P(X \leqslant 1) = 0{\cdot}264.$$

Each of the samples has thus a probability 0·264 of stopping the machine. The number of samples taken between successive stoppages is a geometric variate with a mean $1/0{\cdot}264 = 3{\cdot}79$. Hence the average time interval between successive adjustments is 3·79 hours. □

Problem 6.3 The length of a metal rod produced by a machine can be taken to be a normal variate with standard deviation 0·1 mm. The desired machine setting is 10 mm; and, as a protection against a larger setting, a sample of size n is to be taken every hour from the output of the machine. If the sample means exceeds a limit l, the machine will be carefully checked and set. Find the values of n and l such that the average time interval between successive checks is 100 hours, 10 hours, given that the machine has been set at 10 mm, 10·05 mm respectively.

Solution. Let $\bar{X}$ be the sample mean. Let $P(\bar{X} > l)$ be p_1, p_2 when the machine has been set at the mean levels of 10 mm., 10·05 mm respectively. Thus the mean number of samples taken between successive checks is $1/p_1$, $1/p_2$ if the machine setting is 10·0 mm, 10·05 mm respectively. A sample of n is taken every hour so that we would like $1/p_1 = 100$, $1/p_2 = 10$ or $p_1 = 0{\cdot}01$, $p_2 = 0{\cdot}1$. The sample mean $\bar{X}$ is a normal variate with standard deviation $0{\cdot}1/\sqrt{n}$. Now,

$$P(\bar{X} > l \mid \mu = 10) = 0{\cdot}01 \text{ gives } l = 10 + 0{\cdot}2326/\sqrt{n}, \text{ and}$$

$$P(\bar{X} > l \mid \mu = 10{\cdot}05) = 0{\cdot}1 \text{ gives } l = 10{\cdot}05 + 0{\cdot}1282/\sqrt{n}.$$

Since n is an integer, a reasonable solution is $n = 4$, $l = 10{\cdot}12$. □

Problem 6.4 For the item of problem 2.8 each inspection costs £C. Find the expected cost due to inspection till the detection of failure.

Solution. The inspection times, x_i, $i = 1, 2, 3 \ldots$ are such that

$$P[\text{Failure in } (x_i, x_{i+1}) \mid \text{no failure up to } x_i] = p$$

so that

$$P[\text{no failure in } (x_i, x_{i+1}) | \text{no failure up to } x_i] = 1-p = q.$$

Thus if R is the number of inspections till the detection of failure, we have $P(R = r) = pq^{r-1}$. Since $E(R) = 1/p$, the required expected cost is £C/p. □

Problem 6.5 A physicist has planned a series of experiments involving elementary particles, and the series will end as soon as the first successful experiment is carried out. Each of the experiments has, independently, a probability p of being successful, and the cost of n experiments is £$a+bn$.

Find the expected cost of the series. If the physicist has allocated £k for the series, find the probability that the allocation will be adequate.

Solution. Let N be the number of necessary experiments. Now

$$P(N = n) = (1-p)^{n-1}p, \qquad n = 1.2.3\ldots$$

so that

$$E(\text{cost}) = \sum_{1}^{\infty} (a+bn)(1-p)^{n-1}p$$
$$= a+b/p.$$

Now $a+bx = k$ gives $x = (k-a)/b$. Let n_0 be the largest integer smaller than or equal to x. Thus

$$P\text{ (allocation is adequate)} = P(N \leqslant n_0) = 1-(1-p)^{n_0}.$$ □

A negative binomial distribution arises when a series of independent trials is carried out till a defined event A occurs exactly k times. It is supposed that at each of the trials A has a constant probability p of occurring. The number of necessary trials, R, is a negative binomial variate with

$$P(R = r) = \binom{r-1}{k-1} p^k(1-p)^{r-k}, r=k, k+1,\ldots$$

or equivalently, $P(J = j) = \binom{j+k-1}{k-1} p^k(1-p)^j, j = 0, 1, 2, \ldots,$ where $J = R-k$ is the number of times that A does not occur.

It is interesting to regard R as the sum of k independent geometric variates R_i each with the probability distribution

$$P(R_i = r) = (1-p)^{r-1}p.$$

Since

$$R = R_1+R_2+R_3+\ldots R_k,$$
$$E(R) = \sum E(R_i) = k/p,$$
$$V(R) = \sum V(R_i) = kq/p^2.$$

Also since the probability generating function of a sum of independent variates is the product of the probability generating of the variates, we have

$$E(Z^R) = \pi E(Z^{R_i}) = (pz)^k/(1-qz)^k.$$

Problem 6.6 In an inspection scheme for a large batch of items, the items are classified as either effective or defective. The items are selected at random, one by one, from the batch and the inspection stops as soon as three defectives are observed. If the number of inspected items is twenty or less, the batch is rejected; otherwise it is accepted. Find the probability of accepting a batch containing 5% defective items.

Solution. Because of the size of the batch, we take P(defective items) = 0·05 for each of the items selected for inspection. Let N = number of items inspected. Now

$$P(N = n) = \binom{n-1}{2} 0{\cdot}05^2 \times 0{\cdot}95^{n-3} \times 0{\cdot}05, \qquad n = 3, 4, \ldots$$

$$= \binom{n-1}{2} 0{\cdot}05^3 \times 0{\cdot}95$$

Hence P(accepting a batch) $= 1 - \sum_{3}^{20} P(n) = 0{\cdot}295$. □

Problem 6.7 The number of accidents per year for a given bus driver has a Poisson distribution with parameter M where M itself is a random variable with p.d.f. $\lambda(\lambda m)^{K-1}e^{-\lambda m}/(K-1)!$, $m \geqslant 0$ (K an integer). Show that the distribution of the number of accidents per year for a bus driver chosen at random is negative binomial.

Solution. Let X be the r.v. representing the number of accidents per year for a bus driver chosen at random. Now

$$P(X = r \mid M = m) = e^{-m}.m^r/r!$$

The unconditional probability is obtained by multiplying by the p.d.f. of M and integrating over the possible values of M, i.e.

$$P(X = r) = \int_0^\infty e^{-m}.\frac{m^r}{r!}.\frac{1}{(K-1)!}.\lambda(\lambda m)^{K-1}e^{-\lambda m}\,dm$$

$$= \frac{\lambda^K}{r!(K-1)!}\int_0^\infty m^{r+K-1}e^{-(1+\lambda)m}\,dm$$

$$= \frac{\lambda^K}{r!(K-1)!}.\frac{1}{(1+\lambda)^{r+K}}\int_0^\infty u^{r+K-1}e^{-u}\,du$$

$$= \frac{(r+K-1)!}{r!(K-1)!}\left(\frac{\lambda}{1+\lambda}\right)^K\left(\frac{1}{1+\lambda}\right)^r, \qquad r = 0, 1, 2, \ldots$$ □

Problem 6.8 An inspection scheme for judging the quality of a large batch of components is required. The components will be inspected, at random, one by one and the inspection will stop as soon as d components are found to be defective. The batch will be accepted if the number of components inspected is n or larger. Find n, d such that the probabilities of accepting the batch given that the batch contains 1%, 3% defective components are 0·95, 0·10 respectively. Find the average number of items inspected and compare this with the single sampling scheme of Problem 4.18.

Solution. Let $P_a(p)$ be the probability of accepting a batch with 100p% defective components. Since the batch is large, we can suppose that each of the inspected components has a probability p of being defective. Suppose that defective number $(c+1)$ is encountered when i components have been inspected. A direct argument for $P_a(p)$ is

$$P_a(p) = \sum_{i=n}^{\infty} \binom{i-1}{d} p^d(1-p)^{i-1-d}.$$

Another argument is that the number of components inspected will be n or more, if in n inspected components $(d-1)$ or less defective components are found. So that

$$P_a(p) = \sum_{r=0}^{d-1} \binom{n}{r} p^r(1-p)^{n-r}.$$

We would like n, d to be such that $P_a(0{\cdot}01) = 0{\cdot}95$ and $P_a(0{\cdot}03) = 0{\cdot}10$. Assuming that the Poisson approximation for $\binom{n}{r} p^r(1-p)^{n-r}$ can be used, we see from Problem 4.18 that $n = 400$, and $d = 8$. The expected number of components inspected is d/p, so that for a batch with 1% defective components we would expect to inspect 800 components and for a batch with 3% defective components we would expect to inspect 267 components. For the single sampling scheme of Problem 4.18, the number of components inspected is constant at 400 and it is interesting to note that for batches with 1% defective components, the single sampling scheme requires less inspection (on average); for batches with 3% defective components the single sampling scheme requires more inspection. □

6.2 Multinomial distribution. In a binomial situation we have a series of independent trials and at each of the trials there are only two possibilities, say success or failure; the probability of success is constant at each of the trials. A generalisation of a binomial distribution is the *multinomial distribution* which arises when we have a series of n independent trials each of which may result in one of k possibilities i with probabilities

p_i, $i = 1, 2, 3 \ldots k$. The joint probability of exactly r_i trials resulting in the possibilities i is

$$P(r_1, r_2, r_3 \ldots r_k) = \frac{n!}{r_1!r_2!\ldots r_k!} p_1^{r_1} p_2^{r_2} \ldots p_k^{r_k},$$

where $\sum_i r_i = n$ and $\sum_i p_i = 1$. The name multinomial is an appropriate one for the multinomial probabilities $P(r_1, r_2 \ldots r_k)$ arise from the expansion of $(\sum p_i)^n$. If the random variable R_i denote the number of times the possibilities i result in the series of n trials, the marginal distribution of a subset of the k random variables is also multinomial. For example

$$P(R_1 = r_1, R_2 = r, R_3 = r_3) = \frac{n!}{r_1!r_2!r_3!(n-r_1-r_2-r_3)!} \times p_1^{r_1} p_2^{r_2} p_3^{r_3} (1-p_1-p_2-p_3)^{n-r_1-r_2-r_3},$$

$$P(R_i = r_i) = \binom{n}{r_i} p_i^{r_i} (1-p_i)^{n-r_i}.$$

Note that in the particular case of a single random variable R_i, the marginal distribution is binomial.

The conditional distribution is also multinomial. For example the probability that $(R_1 = r_1, R_2 = r_2, R_3 = r_3/R_j = r_j)$ for $j = 4, 5, \ldots k$ is

$$\frac{(n-s)!}{r_1!r_2!r_3!} (p_1/p)^{r_1} (p_2/p)^{r_2} (p_3/p)^{r_3}$$

where

$$s = \sum_{j=4}^{k} r_j \text{ and } p = p_1 + p_2 + p_3.$$

It is interesting that the information that $R_j = r_j$, $j = 4, 5 \ldots k$ enters the conditional probability quite simply as $\sum_j r_j$.

The joint probability generating function, $E[z_1^{R_1} z_1^{R_2} z_3^{R_3} \ldots z_k^{R_k}]$, is

$$G(z_1, z_2, z_3, \ldots z_k) = (\sum_{i=1}^{k} p_i z_i)^n.$$

Partial derivatives of G, evaluated at $z_i = 1$ will yield various moments, for example

$$\frac{\partial G(1, 1, 1, \ldots 1)}{\partial z_i} = E(R_i) = np_i,$$

as expected since R_i is a binomial variate.

$$\frac{\partial G(1, 1, 1, \ldots 1)}{\partial z_i z_j} = E(R_i R_j) = n(n-1)p_i p_j,$$

so that $C(R_i, R_j) = E(R_iR_j) - E(R_i)E(R_j) = -np_ip_j$; correlation coefficient of R_i, R_j is given by

$$\rho = C(R_i, R_j)/\sqrt{\surd(R_i)(R_j)} = -\sqrt{p_ip_j/(1-p_i)(1-p_j)}$$

Problem 6.9 Find $E(R_i \mid R_j = r_j)$ in the case of a multinomial distribution with parameters $n, p_1, p_2 \ldots p_k$.

Solution. The joint probability, $P(r_2, r_3 \ldots r_k \mid R_j = r_j)$ is given by

$$P(r_2, r_3, \ldots r_i \ldots, r_k \mid r_1) = \frac{(n-r_1)!}{r_2!r_3!\ldots r_k!}(p_2/p-p_1)^{r_2}\ldots(p_k/1-p_2)^{r_k}$$

so that $E(R_i \mid R_1 = r_1) = (n-r_1)(p_i/1-p_1)$.
Note that $E(R_i \mid R_j = r_j)$ is a linear function of r_j. □

Problem 6.10 One classification of the blood of human beings has four classes, *O*, *A*, *B* and *AB*. For a given race, it is known that the proportions belonging to these four classes are p_1, p_2, p_3 and p_4. In a random sample of *n* people from this race, find the probability that the distribution over the four classes will be n_1, n_2, n_3 and n_4 persons respectively. How many people of blood group *O* would one expect to find in the sample? In an investigation people are selected at random, one by one, for blood testing, and the investigation will end as soon as one person of group *AB* is found. Find the expected number of the persons whose blood will be tested.

Solution. We have a multinomial distribution with four classes and $P(n_1, n_2, n_3, n_4)$ is equal to

$$\frac{n!}{n_1!n_2!n_3!n_4!}p_1^{n_1}p_2^{n_2}p_3^{n_3}p_4^{n_4},$$

so that E (number of O group) $= np_1$.
P(testing r people till the first AB) $= (1-p_4)^{r-1}p_4$, $r \geqslant 1$, so that the expected number of people tested in the investigation is $1/p_4$. □

Problem 6.11 Side-effects of an antibiotic are classified as nil, minor, major and unacceptable. For a given class of patient, the probabilities ot observing those side effects in a patient are 0·8, 0·1, 0·008 and 0·092 respectively. Find the probability that in a group of 20 patients,

(i) exactly 10 will have side-effects of class nil,
(ii) the number of patients in the four classes will be 15, 2, 2 and 1.

Solution. For (i) we are interested in the class nil, the rest of the three classes being taken together. Thus we have a binomial distribution with

$n = 20$, $p = 0{\cdot}8$, and the required probability is

$$\binom{20}{10}(0{\cdot}8)^{10}(0{\cdot}2)^{10} = 0{\cdot}002.$$

For (ii) the required probability P is given by

$$P = \frac{20!}{15!2!2!1!}(0{\cdot}8)^{15}(0{\cdot}1)^2\ (0{\cdot}008)^2(0{\cdot}092)^2 = 0{\cdot}002 \qquad \square$$

6.3 Exponential, Gamma, and Weibull distributions A summary of these three distributions is given below.

	Exponential	*Gamma*	*Weibull*
$f(x)$	$\lambda e^{-\lambda x}$, $x > 0, \lambda > 0$	$\lambda^\theta x^{\theta-1}e^{-\lambda x}/\Gamma(\theta)$, $x > 0, \theta > 0, \lambda > 0$	$(\beta/\alpha)(x/\alpha)^{\beta-1}e^{-(x/\alpha)\beta}$, $x > 0, \alpha > 0, \beta > 0.$
$E(X)$	$1/\lambda$	θ/λ	$\alpha\Gamma\left(\frac{1+\beta}{\beta}\right)$
$V(X)$	$1/\lambda^2$	θ/λ^2	$\alpha^2\left[\Gamma\left(\frac{2+\beta}{\beta}\right)-\left\{\Gamma\left(\frac{1+\beta}{\beta}\right)\right\}^2\right]$

An interesting property of an exponential distribution is given in Problem 2.4. The mathematical simplicity of an exponential distribution often results in its use as an approximation to the unknown probability properties in a variety of situations.

The moment generating function of an exponential variate is $(1-t/\lambda)^{-1}$. It follows that if X_i, $i = 1, 2, \ldots K$ are independent random variables and each has the exponential distribution with the same parameter λ, then the moment generating function of $Y = \sum X_i$ is $(1-t/\lambda)^{-K}$. Now the moment generating function of a Gamma variate is $(1-t/\lambda)^{-\theta}$, so that the sum of K independent exponential variates is a Gamma variate; this special way in which a Gamma variate arises as a result of summing independent exponential variates means that θ is a positive integer (In general θ need not be an integer and this special case is often labelled as an Erlang variate). Note that if we put $\theta = 1$ in the probability density function of the Gamma variate, we get the density function of the exponential variate.

Note that if we put $\beta = 1$ in the probability density function of the Weibull variate, we get the exponential distribution again. To see another connection between an exponential and a Weilbull distribution, consider the probability that an exponential variate will exceed x. Now

$$P(X > x) = 1-F(x) = \int_x^\infty \lambda e^{-\lambda t}dt = e^{-\lambda x}, \text{ so that}$$

$$P[\alpha(\lambda X)^{1/\beta} > \alpha(\lambda x)^{1/\beta}] = P(X > x) = e^{-\lambda x}.$$

If now we write $Y = \alpha(\lambda X)^{1/\beta}$, and $y = \alpha(\lambda x)^{1/\beta}$ so that $\lambda x = (y/\lambda)^{\beta}$, we have

$$P(Y > y) = e^{-(y/\alpha)\beta} = 1 - F(y).$$

The probability density function of Y is, by differentiation of $1-F(y)$, $f(y) = (\beta/\lambda)(y/\lambda)^{\beta-1}e^{-(y/\alpha)\beta}$. That is Y is a Weibull variate.

Problem 6.12 Assume that the length of time (in minutes) of a telephone conversation is a random variable with p.d.f. $f(x) = \frac{1}{3}e^{-\frac{1}{3}x}$; $x \geqslant 0$. Find the probability that (a) a call will last at least 3 minutes, (b) a call which has already lasted 3 minutes will last for at least another 3 minutes.

Solution. (a) We require $P(X \geqslant 3)$, where X represents the length of the conversation in minutes

$$p(X \geqslant 3) = \int_3^{\infty} \tfrac{1}{3}e^{-\frac{1}{3}x}\,dx = [-e^{-\frac{1}{3}x}]_3^{\infty} = e^{-1}$$

(b) $P(X \geqslant 6/X \geqslant 3) = P(X \geqslant 6)/P(X \geqslant 3)$

$$= [e^{-\frac{1}{3}x}]_6^{\infty}/[e^{-\frac{1}{3}x}]_3^{\infty} = e^{-1}.$$

Problem 6.13 The time to failure T of a component is a r.v. with p.d.f. $\lambda e^{-\lambda t}, t \geqslant 0$. Show that the conditional probability of failure in the interval (T_0, T_0+h), given that failure has not occurred by time T_0, is independent of T_0.

Solution.

$$\begin{aligned} P &= P(T_0 \leqslant T < T_0+h \mid T > T_0) \\ &= P(T_0 \leqslant T < T_0+h)/P(T > T_0) \\ &= \int_{T_0}^{T_0+h} \lambda e^{-\lambda t}\,dt \Big/ \int_{T_0}^{\infty} \lambda e^{-\lambda t}\,dt \\ &= [-e^{-\lambda t}]_{T_0}^{T_0+h}/[-e^{-\lambda t}]_{T_0}^{\infty} \\ &= [e^{-\lambda(T_0)} - e^{-\lambda(T_0+h)}]/e^{-\lambda T_0} = (1-e^{-\lambda h}). \end{aligned}$$

Note: The 'lack of memory' is a special characteristic of an exponential distribution. □

Problem 6.14 The life time of an item is a random variable T with probability density function $f(t) = \lambda e^{-\lambda t}$, where $t > 0$ and λ is unknown. In an experiment n items are placed on a life test and the items are tested at time τ from the start of the experiment. If there are no failures, or if there are r or more failures in $(0, \tau)$, no further testing is done. No failures indicates a satisfactory value of λ, and r or more failures mean an unsatisfactory value of λ. If the number of failures in $(0, \tau)$ is in the range $(1, r-1)$, the failed items are replaced by good ones and the total of n items are again checked after a further time interval of τ. If now the total number of failures recorded in $(0, 2\tau)$ is c or less, λ is judged to be satisfactory; otherwise it is judged to be unsatisfactory. Find the probability that λ will be judged satisfactory if the true unknown value of λ is λ_0. What is the expected duration of the test?

Solution. $P[\text{failure in } (0,\tau)] = \int_0^\tau \lambda_0 e^{-\lambda_0 t}\, dt = 1 - e^{-\lambda_0 \tau} = p.$
Let $q = 1-p$. Now $P[\text{failure in } (\tau, 2\tau) | T > \tau] = p.$
Let $P(\text{decision that } \lambda \text{ is satisfactory}) = P$. We now have

$$P = P[\text{no failures in } (0,\tau)]$$

$$+\sum_{i=1}^{r-1} P[i \text{ failures in } (0,\tau)^-]P[c-i \text{ or less failures in } (\tau, 2\tau)]$$

$$= q^n + \sum_1^{r-1} \binom{n}{i} p^i q^{n-i} \sum_0^{c-i} \binom{n}{j} p^j q^{n-j}.$$

The duration of test is a random variable which takes the values τ and 2τ.

$$P(\tau) = q^n + \sum_r^n \binom{n}{k} p^k q^{n-k} = p_1 + p_2$$

$$P(2\tau) = \sum_1^{r-1} \binom{n}{i} p^i q^{n-i} = p_3 \cdot$$

Note that $p_1 + p_2 + p_3 = 1.$

$$E(\text{duration}) = (p_1+p_2)\tau + p_3(\tau+\tau) = \tau(1+p_3).$$ □

Problem 6.15 A patient has been asked to take K tests. The time that the patient will spend in taking the test i is a random variable T_i, $i = 1, 2 \ldots K$. These K random variables are independent and each of the T_i is the exponential variate with the same parameter λ. Find the probability that the customer will spend more than d units of time in taking the K tests.

Solution. The total time $T = \sum_{i=1}^{K} T_i$ is a Gamma variate and the probability that T is greater than d is $\int^\infty \frac{\lambda^K x^{K-1}}{(K-1)!} e^{-\lambda x}\, dx$. If we call this probability P_K, then using $u = \lambda x$, we have

$$P_K = \int_{\lambda d}^{\infty} \frac{u^{K-1} e^{-u}}{(K-1)!} = K \frac{e^{-\lambda d}(\lambda d)^{K-1}}{(K-1)!} + P_{K-1}.$$

Repeated substitution gives

$$P_K = e^{-\lambda d} \sum_{r=0}^{K-1} (\lambda d)^r / r!.$$

Note that P_K is the Poisson cumulative distribution function. □

Problem 6.16 Experience has shown that a Weibull random variate is a useful model, particularly for explaining the variation in the 'life time' of items. Find $P(X > x)$ and $E(X^r)$. Compute the age-specific failure rate $r(x)$ and examine the influence of β.

If X, in thousands of operations, is the number of times a micro-switch will operate before failure and if X is a Weibull variate with $\beta = 2$ and $\alpha = 100$, what number of operations can be quoted as a 'guaranteed number' such that the probability of meeting the guarantee is 0·99?

Solution.

$$1-F(x) = \int_x^\infty f(t)\,dt = \int_{(x/\alpha)^\beta}^\infty e^{-u}\,du = \exp-\left(\frac{x}{\alpha}\right)^\beta$$

$$E(X^r) = \int_0^\infty \beta\frac{x^r x^{\beta-1}}{\alpha^\beta}\exp-\left(\frac{x}{\alpha}\right)^\beta dx$$

$$= \int_0^\infty \alpha^r t^{r/\beta} e^{-t}\,dt = \alpha^r\Gamma\left(\frac{r+\beta}{\beta}\right)$$

$$r(x) = f(x)/(1-F(x)) = \beta\, x^{\beta-1}/\alpha^\beta.$$

Thus $r(x)$ is decreasing if $0 < \beta < 1$, $r(x)$ is constant if $\beta = 1$ and $r(x)$ is increasing if $\beta > 1$. For the micro-switch we have

$$1-F(x) = \exp-(0{\cdot}01x)^2$$

We want to determine a guarantee time x_0, such that $P[\text{time to failure} \geqslant x_0] = 0{\cdot}99$. That is we have to solve $\exp-(0{\cdot}01x)^2 = 0{\cdot}99$. Thus the guarantee is about 10 000 operations. □

EXERCISES

1. A random variable X is said to be a uniform variate if its probability density function $f(x) = 1/(b-a)$, $a < x < b$. Find (*i*) the mean, (*ii*) the variance, and (*iii*) the moment generating function of a uniform variate.

2. Peter has evolved a technique of producing antique metal figures which are supposed to belong to the time period (1800, 1900) at a cost of £2.00 per figure. The "antiquity effect" varies from figure to figure and Peter feels that a reasonable model is to take the "age" of a figure to be a uniform variate in the range (1800, 1900). Peter sells a figure in the range (1800, 1825) for £36·00, a figure in the range (1825, 1875) for £20·00, and a figure in the range (1875, 1900) for £8·00. Find Peter's expected profit per figure.

3. The conditional probability density of a random variable X, given that another random variable Y has taken the value y is $f(x|y) = y e^{-xy}$, $x > 0, y > 0$. The marginal probability density function of Y is $g(y) = y^{\theta-1} e^{-y}/\Gamma(\theta)$, $\theta > 1$. Find $P(X > x)$ and $E(X)$.

Note: X turns out to be a Pareto variate named after Professor W. Pareto who used it for describing the distribution of the income of a population.

4. A random variable X has the probability density function

$f(x) = p\lambda_1 e^{-\lambda_1 x} + (1-p)\lambda_2 e^{-\lambda_2 x}, 0 < p < 1, \lambda_1 > 0, \lambda_2 > 0, x > 0.$

Find the mean and variance of X.

5. Peter has lunch at a counter-service cafeteria and finds that with probability p he is the first one at the counter so that he has not to wait in a queue before being served. A reasonable model for Peter's waiting time T is to take T to be a random variable with

$$P(T = 0) = p \text{ and } P(T > t) = (1 - p)e^{-\theta t}, t > 0.$$

Find (i) Peter's expected waiting time, (ii) Peter's expected waiting time given that he is not the first one at the counter.

6. A random variable X is said to be a Beta variate if its probability density function is $f(x) = \dfrac{1}{B(m, n)} x^{m-1}(1-x)^{n-1}$, where $m > 0$, $n > 0$, $0 < x < 1$, $B(m, n) = \int_0^1 x^{m-1}(1-x)^{n-1}\,dx$. Find the mean and variance of a Beta variate.

Answers to Exercises

Chapter 1

1. $\frac{3}{8}$.
2. $\sum_{i=0}^{s-1} \binom{i+r-1}{i} p^r q^i$.
3. $\frac{11}{12}$.
4. $\frac{1}{k}\left[1-\left(-\frac{1}{k-1}\right)^{n-1}\right]$.
5. (a) 0·8495, (b) 0·1315.

Chapter 2

1. (i) $0{\cdot}01+0{\cdot}02r$, (ii) 6·15, (iii) $a^2/18$.
2. (i) $3/(1-e^{-10^6})$, (ii) $2p^2q$, where $p = (1-e^{-60^3})/(1-e^{-10^6})$, $q = 1-p$, (iii) $q^2(1+2p)$.
3. $(k+1/\Omega, \Omega^{-2})$.
4. Chebyshev's bound is $\frac{1}{2}$, which is wide in relation to the exact value of $\frac{1}{9}$.
5. $\frac{17}{36}$.
6. $\frac{8}{a^2t^2}\left[1+\left(\exp\frac{at}{2}\right)\left(\frac{at}{2}-1\right)\right]$.

Chapter 3

1. (i) $\frac{1}{3}$, (ii) $\frac{4}{5}$, (iii) $\frac{24}{25}$.
2. $(1-e^{-1})/(1-e^{-1}+e^{-2})^{\frac{1}{2}}$.
3. $\frac{2}{3}$
4. $\binom{k}{r} B(r+m, k+n-r)/B(m, n)$.
5. (i) $G\{H(z)\}$, (ii) $\mu\mu_N$, (iii) $\mu^2\sigma_N^2+\sigma^2\mu_N$.

Chapter 4

1. $6p^2(1-p)^2$, where $p = (\frac{1}{2})^{m+1}$.
2. 0·787.

3. $\dfrac{\pi p_1^n}{1-(1-p_1)^n}+\dfrac{(1-\pi)p_2^n}{1-(1-p_2)^n}$

4. $4\theta(2-\theta)(1-\theta)^6$

5. $1-e^{-1}$.

6. $\dfrac{(\lambda p)^k e^{-\lambda p}}{k!}$

Chapter 5

1. 10·54 and 0·054
2. 0·0127.
4. 0·998

Chapter 6

1. (i) $(a+b)/2$, (ii) $(b-a)^2/12$, (iii) $(e^{-bt}-e^{-at})/t(b-a)$.
2. £19·00.
3. $1/(x+1)^\theta$; $1/(1-\theta)$.
4. $p/\lambda_1+(1-p)/\lambda_2$; $p(1-p)(1/\lambda_1-1\lambda_2)^2+p/\lambda_1^2+(1-p)/\lambda_2^2$.
5. (i) $(1-p)/\theta$, (ii) $1/\theta$.
6. $m/(m+n)$, $mn/(m+n)^2(m+n+1)$.

Appendix

TABLE 1. Distribution function of a binomial variate.

The table gives values of 10 000 $F(c) = 10\,000 \sum_{r=0}^{c} \binom{n}{r} p^r (1-p)^{n-r}$ for a selected set of values of n, p, c.

n	c	p 0·05	0·10	0·15	1/6	0·25	0·30	0·35	0·40	0·45	0·50
2	0	9025	8100	7225	6944	5625	4900	4225	3600	3025	2500
	1	9975	9900	9775	9722	9375	9100	8775	8400	7975	7500
4	0	8145	6561	5220	4822	3164	2401	1785	1296	0915	0625
	1	9860	9477	8905	8681	7383	6517	5630	4752	3910	3125
	2	9995	9963	9890	9838	9492	9163	8735	8208	7585	6875
	3	10 000	9999	9995	9992	9961	9919	9850	9744	9590	9375
6	0	7351	5314	3771	3349	1780	1176	0754	0467	0277	0156
	1	9672	8857	7765	7368	5339	4202	3191	2333	1636	1094
	2	9972	9841	9527	9377	8305	7443	6471	5443	4415	3438
	3	9999	9987	9941	9913	9624	9295	8826	8208	7447	6563
	4	10 000	9999	9996	9993	9954	9891	9777	9600	9308	8906
	5		10 000	10 000	10 000	9998	9993	9982	9959	9917	9844
8	0	6634	4305	2725	2326	1001	0476	0319	0168	0084	0039
	1	9428	8131	6571	6047	3671	2553	1691	1064	0632	0352
	2	9942	9619	8948	8651	6785	5518	4278	3154	2201	1445
	3	9996	9950	9786	9693	8862	8059	7064	5941	4770	3633
	4	10 000	9996	9971	9954	9727	9430	8940	8263	7396	6367
	5		10 000	9998	9996	9958	9887	9747	9501	9115	8555
	6			10 000	10 000	9996	9987	9964	9915	9819	9648
	7					10 000	9999	9998	9993	9983	9961
10	0	5987	3487	1969	1615	0563	0282	0135	0051	0025	0010
	1	9139	7361	5443	4845	2450	1493	0859	0406	0233	0107
	2	9885	9298	8202	7752	5256	3828	2616	1517	0996	0547
	3	9990	9872	9500	9303	7759	6496	5138	3575	2660	1719
	4	9999	9984	9901	9845	9219	8597	7515	6078	5044	3770
	5	10 000	9998	9996	9976	9803	9526	9051	8165	7384	6231
	6		10 000	9999	9997	9965	9894	9740	9374	8980	8281
	7			10000	10 000	9996	9984	9952	9854	7736	9453
	8					10 000	9999	9995	9979	9955	9893
	9						10 000	10 000	9999	9997	9990

TABLE 2. Distribution function of a Poisson variate

The table gives values of 10 000 $F(c) = \sum_{r=0}^{c} \lambda^r e^{-\lambda}/r!$ for a selected set of values of λ, c.

	λ									
c	0·05	0·10	0·15	0·20	0·25	0·30	0·35	0·40	0·45	0·50
0	9512	9048	8607	8187	7788	7408	7047	6703	6376	6065
1	9988	9953	9898	9825	9735	9631	9513	9385	9246	9098
2	10 000	9999	9995	9989	9978	9964	9945	9921	9891	9856
3		10 000	10 000	9999	9999	9997	9995	9992	9988	9983
4				10 000	10 000	10 000	10 000	9999	9999	9998
5									10 000	10 000

TABLE 2—(*continued*)

	λ									
c	0·60	0·70	0·80	0·90	1·0	1·5	2·0	2·5	3·0	3·5
0	5488	4966	4493	4066	3679	2231	1353	0821	0498	0302
1	8781	8442	8088	7725	7358	5578	4060	2873	1992	1359
2	9769	9659	9526	9371	9197	8089	6767	5438	4232	3209
3	9964	9943	9909	9865	9810	9344	8571	7576	6472	5366
4	9996	9992	9986	9977	9963	9814	9474	8912	8153	7254
5	10000	9999	9998	9997	9994	9955	9834	9580	9161	8576
6		10000	10000	10000	9999	9991	9955	9858	9665	9347
7					10000	9998	9989	9956	9881	9733
8						10000	9998	9989	9962	9901
9							10000	9997	9989	9967
10								9999	9997	9990
11								10000	9999	9997
12									10000	9999
13										10000

	λ									
c	4	5	6	7	8	9	10	11	13	15
0	0183	0067	0025	0009	0003	0001	0001			
1	0916	0404	0174	0073	0030	0012	0005	0002		
2	2381	1247	0620	0296	0138	0062	0028	0012	0002	
3	4335	2650	1512	0818	0424	0212	0103	0049	0011	0002
4	6288	4405	2851	1730	0996	0550	0293	0151	0037	0009
5	7851	6160	4457	3007	1912	1157	0671	0375	0107	0028
6	8893	7622	6063	4497	3134	2068	1304	0786	0259	0076
7	9489	8666	7440	5987	4530	3239	2202	1432	0540	0180
8	9786	9319	8472	7291	5926	4556	3328	2320	0998	0375
9	9919	9682	9161	8305	7166	5874	4579	3405	1658	0699
10	9972	9863	9574	9015	8159	7060	5830	4599	2517	1185
11	9991	9946	9799	9467	8881	8030	6968	5793	3532	1848
12	9973	9980	9912	9730	9362	8758	7916	6887	4631	2676
13	9999	9993	9964	9872	9658	9262	8645	7813	5731	3632
14	10000	9998	9986	9943	9827	9585	9165	8540	6751	4657
15		9999	9995	9976	9918	9780	9513	9074	7636	5681
16		10000	9998	9990	9963	9889	9730	9441	8355	6641
17			9999	9996	9984	9947	9857	9678	8905	7489
18			10000	9999	9994	9976	9928	9823	9302	8195
19				10000	9998	9989	9966	9907	9573	8752
20					9999	9996	9993	9953	9750	9170
21					10000	9998	9997	9978	9859	9469
22						9999	9999	9990	9924	9673
23						10000	10000	9995	9960	9805
24								9998	9980	9888
25								9999	9990	9938
26								1000	9996	9967
27									9998	9983
28									9999	9991
29									10000	9996
30										9998
31										9999
32										10000

TABLE 3. Poisson parameter λ.

The table gives values of 100λ for a selected set of values of $P(R \leqslant c) = \sum_0^c \lambda^r e^{-\lambda}/r!$

c	$P(R \leqslant c) = \sum_0^c \lambda^r e^{-\lambda}/r!$						
	0·95	0·90	0·75	0·50	0·25	0·10	0·05
0	5	10	29	69	139	230	300
1	35	53	96	168	269	389	474
2	82	110	173	267	392	532	630
3	137	175	254	367	511	668	775
4	197	243	337	467	627	799	915
5	261	315	422	567	742	928	1051
6	329	390	508	667	859	1053	1184
7	398	466	596	767	968	1177	1315
8	470	543	684	867	1080	1300	1443
9	543	622	773	967	1191	1421	1571
10	617	702	862	1067	1302	1541	1696
11	692	783	952	1167	1412	1660	1821
12	769	865	1042	1267	1522	1778	1944
13	846	947	1133	1367	1631	1896	2067
14	925	1030	1224	1467	1740	2013	2189

TABLE 4. Distribution function of a standard normal variate.

The table gives values of $10\,000F(x) = 10\,000\int_{-\infty}^{x} \frac{1}{\sqrt{2\pi}} e^{-t^2/2}\,dt$ for a selected set of values of x.

x	·00	·01	·02	·03	·04	·05	·06	·07	·08	·09
0·0	5000	5040	5080	5120	5160	5199	5239	5279	5319	5359
·1	5398	5438	5478	5517	5557	5596	5636	5675	5714	5753
·2	5793	5832	5871	5910	5948	5987	6026	6064	6103	6141
·3	6179	6217	6255	6293	6331	6368	6406	6443	6480	6517
·4	6554	6591	6628	6664	6700	6736	6772	6808	6844	6879
·5	6915	6950	6985	7019	7054	7088	7123	7157	7190	7224
·6	7257	7291	7324	7357	7389	7422	7454	7486	7517	7549
·7	7580	7611	7642	7673	7703	7734	7764	7793	7823	7852
·8	7881	7910	7939	7967	7995	8023	8051	8078	8106	8133
·9	8159	8186	8212	8238	8264	8289	8315	8340	8365	8389
1·0	8413	8438	8461	8485	8508	8531	8554	8577	8599	8621
1·1	8643	8665	8686	8708	8729	8749	8770	8790	8810	8830
1·2	8849	8869	8888	8906	8925	8943	8962	8980	8997	9015
1·3	9032	9049	9066	9082	9099	9115	9131	9147	9162	9177
1·4	9192	9207	9222	9236	9251	9265	9279	9292	9306	9319
1·5	9332	9345	9357	9370	9382	9394	9406	9418	9429	9441
1·6	9452	9463	9474	9484	9495	9505	9515	9525	9535	9545
1·7	9554	9564	9573	9582	9591	9599	9608	9616	9625	9633
1·8	9641	9648	9656	9664	9671	9678	9686	9693	9699	9706
1·9	9713	9719	9726	9732	9738	9744	9750	9756	9761	9767
2·0	9772	9778	9783	9788	9793	9798	9803	9808	9812	9817

x	·00	·01	·02	·03	·04	·05	·06	·07	·08	·09
2·1	9821	9826	9830	9834	9838	9842	9846	9850	9854	9857
2·2	9861	9864	9868	9871	9875	9878	9881	9884	9887	9890
2·3	9893	9896	9898	9901	9904	9906	9909	9911	9913	9916
2·4	9918	9920	9922	9924	9927	9929	9930	9932	9934	9936
2·5	9938	9940	9941	9943	9945	9946	9948	9949	9951	9952
2·6	9953	9955	9956	9957	9959	9960	9961	9962	9963	9964
2·7	9965	9966	9967	9968	9969	9970	9971	9972	9973	9974
2·8	9974	9975	9976	9977	9977	9978	9979	9979	9980	9981
2·9	9981	9982	9982	9983	9984	9984	9985	9985	9986	9986
3·0	9986	9987	9987	9988	9988	9989	9989	9989	9990	9990
3·1	9990	9991	9991	9991	9992	9992	9992	9992	9993	9993
3·2	9993	9993	9994	9994	9994	9994	9994	9995	9995	9995
3·3	9995	9995	9995	9996	9996	9996	9996	9996	9996	9996
3·4	9997	9997	9997	9997	9997	9997	9997	9997	9997	9998
3·5	9998	9998	9998	9998	9998	9998	9998	9998	9998	9998
3·6	9998	9998	9998	9999	9999	9999	9999	9999	9999	9999

TABLE 5. Percentage points of a standard normal variate.

The table gives values of x for a selected set of values of $P(X > x) = \int_x^\infty \frac{1}{\sqrt{2\pi}} e^{-t^2/2}\,dt$

$P(X > x)$	x	$P(X > x)$	x	$P(X > x)$	x
0·50	0·0000	0·10	1·2816	0·003	2·7478
0·45	0·1257	0·05	1·6449	0·002	2·8782
0·40	0·2533	0·04	1·7507	0·001	3·0902
0·35	0·3853	0·03	1·8808	0·0005	3·2905
0·30	0·5244	0·02	2·0537	0·0004	3·3528
0·25	0·6745	0·01	2·3263	0·0003	3·4316
0·20	0·8416	0·005	2·5758	0·0002	3·5401
0·15	1·0364	0·004	2·6521	0·0001	3·7190